FIND YOUR PATH

FIND YOUR PATH

Unconventional Lessons from 36 Leading
Scientists and Engineers

Daniel Goodman

The MIT Press
Cambridge, Massachusetts
London, England

This book was set in Stone Serif by Westchester Publishing Services. Printed and bound in the United States of America.

Library of Congress Cataloging-in-Publication Data
Names: Goodman, Daniel (Daniel Lawrence).
Title: Find your path : unconventional lessons from 36 leading scientists and
    engineers / Daniel Goodman.
Description: Cambridge, MA : The MIT Press, [2019]
Identifiers: LCCN 2018059708 | ISBN 9780262537544 (pbk. : alk. paper)
Subjects: LCSH: Science--Vocational guidance. | Engineering--Vocational
    guidance. | Scientists--Biography. | Engineers--Biography. | Women
    scientists--Biography. | Women engineers--Biography.
Classification: LCC Q147 .R63 2019 | DDC 502.3--dc23 LC record available at
    https://lccn.loc.gov/2018059708

10 9 8 7 6 5 4 3 2 1

CONTENTS

▌▌ THE ACADEMICS

PREFACE

Scientists and engineers can find challenging, financially successful, and intellectually satisfying careers in a range of academic, business, and government settings. Of these possibilities, the most familiar to students is probably the academic environment. Students spend years studying with professors at colleges and universities and so have some understanding what an academic career might be like. They may be less familiar with the range of career possibilities in industry or government service.

This book was originally conceived to provide *career guidance* to students considering careers in these three settings. However, as the project progressed, I realized that through their first-person stories our role-model scientists and engineers are actually providing valuable *life guidance* for readers at any stage in their career.

An example of life guidance is how to create a career that achieves balance between family life and work. Another example is how best to find a mentor or career champion. Many of the women in this book have experienced gender discrimination at some point in their careers. In their chapters, they describe their thoughts, feelings, and reactions to this discrimination. The choices they made provide excellent examples of life guidance. Some individuals explain their process of incorporating strongly held values into their training or workplace. Our minority subjects have faced a range of challenges, whether due to growing up in an underserved community or to experiencing racial bias or to being a pioneer at their school or workplace.

The personal and professional choices our subjects have made flow from the environment in which they grew up, including the direction and encouragement by parents and teachers as well as from their innate personality, talents, and interests. Although their challenges vary from subject to subject, one commonality is the love these role models have for scientific discovery and their desire to use science and technology to understand and improve the world.

The idea for this book came from my interactions with students and with exceptional scientists through the Fannie & John Hertz Foundation. Every summer the Hertz Foundation holds a workshop where the in-school fellows get to interact with older fellows and hear talks by guest speakers. Some of the guest speakers have started companies, some are university professors working in interesting research areas, and some are government scientists or administrators. At the summer workshop in 2014, Kimberly Budil gave a lovely talk about her career in government service. As a longtime manager at the National Laboratories, Kim described how much she enjoys enabling and motivating teams of scientists. She talked about the challenges and successes of her career, including her experiences as a mother balancing responsibilities for her family with her commitment to her workplace. After her talk, I thought about ways to disseminate the wisdom of role-model scientists such as Kimberly and decided to create a book.

This book is also the result of my experiences with interns and visitors at our company, ASM-NEXX, where I am a director. From time to time, we arrange an event where our scientists tell students about their current work and their career experiences. Hearing personal narratives allows students and visitors to imagine themselves in the scientists' shoes and consider whether they might want to pursue a similar career. As a result, I chose to present these life stories and career guidance using the effective format of the personal narrative.

PROCESS AND ORGANIZATION

The majority of the essays were drafted based on interviews with each subject. Subjects had the opportunity to review and suggest changes to or rewrite their essay. Two scientists (Richard Miles and Richard Post) preferred to write their own first draft. Each essay presents a main theme in its introduction. The narratives are then structured chronologically, following the individual from childhood through early career. Many of the essays have highlighted insert matter in which the subject provides comments or answers questions posed by the interviewer.

The 36 essays are divided into three groups. The first group, "The Entrepreneurs," includes scientists and engineers who have worked in commercial settings. The second group, "The Academics," includes individuals who have spent much of their careers at universities or nongovernmental research institutions. Those in the third group, "The Public Servants," have spent much of their careers working for the US government, either as civilian or military employees.

Some of the subjects in the first group have founded companies; they are entrepreneurs in the most common use of the term. Others in this group have made important technical and leadership contributions at existing enterprises. By creating economic value and growing companies, these men and women are also entrepreneurs in the term's original meaning. Although I have assigned each subject to one of the three groups, some have had careers that span across categories. For example, several entrepreneurs have also spent time in academic and government settings. Likewise, some of the academic scientists have worked for the government or consulted for commercial companies or both.

It has been a pleasure getting to know and write about these wonderful individuals. They have taught me many lessons about life and about the way successful scientists make choices about their training and career. I also learned that it is possible to have many careers during a lifetime and that it is never too late to start chasing your dream. I hope you will find these essays as enjoyable to read as they were to create.

ACKNOWLEDGMENTS

I thank each of the role-model scientists and engineers for agreeing to participate in this project. Our subjects were generous with their time and enthusiasm, and several also made donations to help underwrite the project.

This book is the result of a career-guidance project sponsored by the Fannie & John Hertz Foundation. Since 1963, the Hertz Foundation has been providing PhD fellowships to exceptionally talented individuals expected to have the greatest impact on the application of science, math, and engineering to human problems. The intellectual freedom granted by the fellowship has fostered a cross-generational community of research leaders and entrepreneurs; Hertz fellows include Nobel laureates, CEOs, generals, and best-selling authors. The foundation is the legacy of John Hertz, a Hungarian immigrant who made his fortune by capitalizing on entrepreneurship prospects in the automotive industry and who believed that innovative and entrepreneurial solutions were vital to the strength, security, and prosperity of the United States. For more information on the Hertz Foundation and the cutting-edge innovations led by Hertz Fellows, please visit www.hertzfoundation.org.

I appreciate the assistance of many members of the Hertz Foundation community. The book began from conversations with Emma Rosenfeld. Jennifer Schloss, Ashvin Bashyam, and Samuel Rodriques helped interview subjects and edit essays. Bennett McIntosh assisted with interviewing and essay writing. Jayne Iafrate helped with photo editing. I also thank Robbee Kozak, Richard Miles, and David Galas, who have supported this project from the beginning.

I thank Greta Sibley, who created the first book layout, and my editors at the MIT Press, Jermey Mathews, Deborah Cantor-Adams, and Annie Barva, who guided the project to publication. I also thank Laura Serra, who sketched lovely portraits of each subject, and Christine Palm, who provided copyediting. I also appreciate helpful comments from Sasha Bergmann, Susan Cohen, Eliora Goodman, and Joel Segel.

Finally, I thank my family for their love and support during this project. Many thanks to Lisa for excellent suggestions throughout the project and to Eliora, Seth, and Hannah for providing me with their own set of important life lessons.

Daniel Goodman
Lexington, Massachusetts

HOW TO READ THIS BOOK

One way to read this book is by starting at the beginning, continuing in order through the three groups of essays in order and then reading the summary chapter, "Life and Career Lessons: Learning from Lives, Views, and Stories."

Some readers will find it useful to begin with the summary chapter before reading any of the essays. The summary chapter contains the findings of a fascinating four-year career-guidance research project. It begins with the reasons each of the subjects chose his or her career direction and the traits and preferences shared by members of his or her career group. It then breaks down the essays by the life and career lessons that subgroups have in common. The summary chapter also includes questions for the reader to consider in order to incorporate ideas from these essays into their own lives.

Some readers with particular interests may decide to start by reading a subset of the essays. For example, students may find the essays about subjects who are at a point that is still early in their careers to be most relevant to their current challenges and career decisions. Early-career subjects include *Ethan Perlstein, Christopher Loose, Beth Reid, Z. Jane Wang, Jennifer Park, Jessica Seeliger, Stephon Alexander, Jennifer Roberts, Renee Horton,* and *Jami Valentine.*

Readers who are thinking of joining an early-stage company will enjoy reading the entrepreneurial experiences of *Christopher Loose, Sandra Glucksmann, Richard Lethin, Stephen Fantone, Daniel Theobald, Beth Reid, Z. Jane Wang, Daniel Goodman, David Galas,* and *Richard Post.*

Some of the entrepreneurs became technical leaders or executives at large companies. Readers interested in this career path may want to start with essays about the careers of *Kathy Gisser, Kathleen Fisher, Wanda Austin,* and *Norman Augustine.*

Readers interested in scientific careers related to national defense can start by reading essays about the lives of *Wanda Austin, Norman Augustine, Michael O'Hanlon, William Press, Richard Miles, Jennifer Roberts, Kimberly Budil, Wendy Cieslak, Ellen Pawlikowski, Paul Nielsen,* and *Jay Davis.*

Readers interested in nondefense research careers in the public sector may want to begin by reading the experiences of *Renee Horton, Jami Valentine, Ellen Stofan, John Mather, Jay Davis,* and *David Galas.*

Several of the subjects changed scientific fields at various stages in their training or careers. Readers who are thinking about changing fields will be interested in essays about *Michael O'Hanlon, William Press, Shirley Tilghman,* and *David Galas.*

Many of the scientists and engineers found mentors early in their training to be crucial to their careers. Readers interested in this topic may want to start with essays about the experiences of *Kathy Gisser, Jessica Seeliger, Tamara Doering,* and *Rainer Weiss.*

Readers interested in issues of work–life balance will enjoy reading essays about the lives of *Sandra Glucksmann, Daniel Theobald, Daniel Goodman, Wanda Austin, Jessica Seeliger, Tamara Doering, Jennifer Roberts, Jami Valentine, Deirdre Olynick,* and *Ellen Pawlikowski.*

Many of the women scientists have encountered organizational barriers, gender bias, or discrimination during their careers. Minority subjects in this book have had to overcome a wide range of societal challenges to become successful scientists and leaders. As a result, the proportion of women and minorities in science and engineering in the United States remains stubbornly low. Regardless of how you choose to read this book, I encourage you to include in your selection the exceptional experiences of *Deirdre Olynick, Jennifer Roberts, Wendy Cieslak, Wanda Austin, Stephon Alexander, Renee Horton,* and *Jami Valentine.* They are truly role-model scientists and engineers with life and career lessons worth learning.

THE ENTREPRENEURS

1 LOVE OF THE UNKNOWN

Sandra Glucksmann

A start-up company is like dating and puppy love—everything is new. Of course, you get to learn about a new area of science, and you get to make decisions about all aspects of the business—agreeing on business and scientific strategies and plans, building a team, designing a lab. You get to decide what expertise you want to bring in, what kind of culture you want to develop, and how you are going to raise money. It is like multitasking to the nth degree. For me, there is nothing more thrilling. It is challenging in so many ways: exhausting and totally time consuming. It is irresistibly compelling, and I love it.

I was born in Argentina, and from the time I was eight until I was 24 and started graduate school, I hadn't lived in the same country for more than three years in a row. During that time, I attended six different elementary schools, one middle school, three high schools, and two colleges all over the United States and Latin America. Such a childhood prepared me to be comfortable with the uncertainty that accompanies entrepreneurship. That's what the leader of my previous company said when she learned about my background; she understood better why I thrive in a start-up world: I like the unknown—it doesn't scare me.

It's actually the known that scares me. When I start feeling too comfortable with what I am doing, it creates a lot of anxiety. That feeling was one of the reasons I left the first company I worked for, where I had risen to vice president in charge of several hundred employees, to start my own company. And now I have left that company for yet another venture.

SHUTTLING BETWEEN COUNTRIES

My parents got divorced when I was five, and both remarried. I grew up in a household with my mother and my stepfather and moved with them between countries as my stepfather pursued his career. In the 1960s, when he was in charge of public relations and strategy for Ford Motor Company, we moved to the United States and lived in Michigan. Then our family moved back to Argentina, where my stepdad worked for a scientific foundation in Patagonia. When the politics in Argentina got bad, my mother, my four siblings, and I again moved to the United States.

My stepfather, Julio, was an important influence while I was growing up. Julio was a multitalented person, a true Renaissance man, with an MBA from Berkeley and a political science degree from the Sorbonne. He taught me and my siblings many things, including how to appreciate nature and art and how to understand politics. He helped make us aware of our place in the world and how we could make a difference.

I grew up in the late 1960s and early 1970s, a time when ecology and environmental sustainability were prominent topics of discussion. There was a fear that we wouldn't be able to feed the world's growing population. I became very aware of the environment and what humans were doing to it. For example, in Argentina everyone ate meat. Even before I became a vegetarian, I was aware that eating meat was probably not good for the environment. I would lecture my older brother, telling him not to eat too much meat. I was a bossy 10-year-old.

Like my stepfather, my mother is also multitalented. She can cook a wonderful meal, has started multiple restaurants, and speaks six languages. Even though she has no formal training, she can design a house, laying it out like an architect. She comes from a generation of women who didn't necessarily have the opportunity to go to college. She always wanted to be a chemist but married when she was young, had five kids, and was divorced by age 27. Taking care of her family took precedence over getting a college degree.

I grew up always being a good girl and trying to please. Since my mother always wanted to be a scientist, that felt like the right thing for me to do; I knew it would please my parents. Throughout high school, I loved science and math. Although I also liked chemistry and biochemistry, the subject that made the most sense to me was biology. It seemed rational. Biology takes the basic concepts of chemistry and physics, such as entropy, and applies them to daily life and nature. It was a science to which I could relate. Biology helps me to navigate the world and feel more comfortable in it.

When I finished high school in the United States, I wrote to my father, who was working in Argentina, and told him that I was considering going to college to study biology. But I was also thinking of other possibilities, perhaps studying a social science such as anthropology or perhaps even literature. He wrote back and suggested that I come back to Argentina, spend a year, and figure it out. I was 17 at the time. When I arrived, I found out that my high school diploma from the United States wasn't recognized in Argentina and that I needed to complete a baccalaureate degree, so I finished high school again and then applied to college.

AN EARLY SPECIALIZATION

I spent my first two years of college at the Catholic University in Buenos Aires. Higher education in Argentina is similar to the European system—each college teaches a specialized subjects such as biology, agriculture, medicine, or law. I entered the College of Biology, where I also studied statistics, physics, and chemistry. And since it was a Catholic school, I also had to study philosophy, but not religion.

During the summers, I visited my mom and stepfather, who were living near Chicago. My stepfather suggested that I transfer to a US college, saying that if I wanted a career in applied science, it would be better to study and live in a country with more research opportunities. Julio was working for an international seed company at the time and collaborating with a botanist at the University of Wisconsin. We went to visit his collaborator and also toured the campus. In 1980, the application and acceptance process were less formal than today. When we discussed the possibility that I might transfer, the Department of Biology said they would accept me for enrollment in September and that I should send in my transcripts.

The University of Wisconsin gave me transfer credit for all the courses I had taken in Argentina. Even though I had been in college in Argentina for only two years, I had already taken all the core undergraduate classes, which allowed me to take graduate-level courses such as virology and molecular biology during my two years at Wisconsin. I also did research in a molecular biology lab working on transcriptional regulation using bacteria and virus as models to understand cellular molecular machinery. From this research and my classes, I decided I wanted to become a molecular geneticist. I applied to graduate schools in the Chicago area because I was about to get married to a lawyer in Chicago. I graduated in December, and we got married in February. I was accepted in the Department of Molecular Genetics and Cell Biology at the University of Chicago but didn't want to wait until September to get back into my studies, so I started working in a research lab in the summer. The lab in which I did my first rotation was the same one in which I ultimately did my PhD thesis.

A LIFELONG INFLUENCE

As part of the discussions I had with various professors when I started at Chicago, I met Professor Lucia Rothman-Denes. I got along really well with Lucia. We had similar backgrounds: both of us were born in Argentina and had a Jewish father and a Catholic mother. We had similar personalities, and even though I was considerably younger, we had a similar appearance, with very short salt-and-pepper hair.

Lucia and I are still close friends. She is a person who has truly influenced me and been a wonderful mentor. I admire her personally, and she has influenced me scientifically—from Lucia, I learned about the rigor of science. She was a very demanding adviser, but she was also a wonderful mother, a wonderful wife, and now a grandmother. She lives her life very fully, and I appreciate having met her and having her as an important part of my life.

My first semester at Chicago, I took Lucia's foundational class, which made me excited to work in molecular genetics. In her lab, we worked on an interesting virus-model system, studying the relationship between protein structure and function. If we made a change to a protein or DNA molecule, within a matter of hours we could see the effect on the virus and on the function of that protein. Being able to see quickly the results of an experiment is something I love about molecular biology.

For my thesis, I worked on the transcriptional regulation of this model virus, which has an RNA polymerase that reads the DNA to make RNA. The polymerase is an unusually large protein and so in itself is an interesting molecule to study. But how that protein recognizes DNA is also very unusual. I was able to create mutations at the DNA level and see the effect those mutations had on the protein's ability to carry out its function. The interesting part wasn't just the effect the mutations had, but the effect the mutations had on secondary structure formed at the DNA level. I showed that this secondary structure was more important than the actual nucleic acid sequence.

Although our findings were generally consistent with the understanding of transcriptional regulation, they were not predictable, which is what made the work interesting. Even though I like the predictability and rational nature of biology, the beauty of my research in graduate school was finding something new every day.

JUGGLING GRAD SCHOOL AND FAMILY

My husband and I lived in Hyde Park during the five and a half years it took me to complete my PhD. We had a daughter my third year in graduate school—not something we had planned, but the love of our life. This new factor also made life interesting—being in graduate research, married to a litigator, and having a toddler. I don't think having a toddler affected my graduate school career that much, but it did have a big effect on our work–life balance and our marriage. Charlie and I lost most of the liberties we had before our daughter was born. If he had a trial or if I had to go back to the lab to do an experiment, we needed to juggle our schedules to make it work.

When I was finishing graduate school and deciding on where to do a post-doc, location was an important consideration. I wanted to live in a city where there would be academic job possibilities after the postdoc. We wanted to move to a state where Charlie could take the bar exam and not have to move yet again. And we wanted a city on the East Coast to be close to our families. Cambridge, Massachusetts, met these requirements and was our top choice.

With my long-standing love of the environment, I wanted to apply to plants some tools of molecular genetics I had learned in graduate school. I looked at labs that were working on plant molecular biology and ultimately went to Graham Walker's lab at MIT to work on symbiosis and relationships between bacteria and plants.

The research during my postdoc at MIT progressed well enough, but I didn't like the academic environment nearly as much as what I had experienced in Chicago. In Lucia's lab, I had been on a project that no one else in the world was working on. My thesis work led to significant results and was published as two papers in a top-tier journal. During graduate school at Chicago, I never needed to deal with the politics and competition of academic science. So when I arrived in Cambridge, the competitive environment at Harvard and MIT was a shock. I felt naive about the politics of science—the need to be better than the next scientist seemed so cutthroat.

I called Lucia on the phone to talk about the direction of my career and my mixed feelings about the postdoc. I told her that I had moved to Cambridge thinking it would be an academic heaven at the center of the research world, but in reality the researchers were overly competitive, not any smarter or the work any better than in her lab, and I did not feel as intellectually challenged as I had felt in her lab. Lucia was sympathetic and reminded me that the main part of my scientific growth had already happened during graduate school.

As an adviser, Lucia had been an incredibly demanding, which I think is typical of women scientists mentoring other women. In Lucia's lab, even though I had first-author publications, I never felt I was doing well enough. It was only after I left her lab that she said to me, "Remember, Sandra, no matter what you set out to do, you will do well." Hearing those words, I asked myself why she had never said them to me when I was in her lab. Lucia mentored graduate students from other labs, encouraging them and helping them think about their careers. But in her own lab, she only discussed how to do great science. I think the reason may have been that she considered her students a reflection of herself, as if she were our mother, and she was as demanding of her students as she was of herself.

AN UNEXPECTED DIRECTION

The entrepreneurial direction of my career after my postdoc happened by chance. I had been anticipating a career in a university setting, especially in the department at the University of Chicago where I had trained, and 27 years ago researchers seldom left academia to join start-up companies.

When I was finishing my postdoc, however, one of my friends told me that he had joined a start-up called Millennium Pharmaceutical. He asked me for my CV, which led to an interview with Eric Lander, one of the company founders. I asked Graham Walker his opinion of the opportunity, worried that he would think less of me if I left academia. But he was supportive and offered to provide a recommendation. A few weeks later I interviewed again with the small group in the company. I was quite impressed with the company and its goals. In the early 1990s, we were at the start of the genomic revolution and beginning to have the tools to be able to sequence DNA and understand genetic components of complex diseases such as cancer, inflammation, diabetes, and obesity. I had always been drawn to the practical uses of science and thought these goals very exciting. I joined Millennium as one of its first scientists.

My first job was in the Genomics Department as a physical mapper: cloning to identify the mutation that causes particular diseases. My team and I identified the gene for polycystic kidney disease, a rare disorder in which cysts grow in the kidneys. I then led a team that identified the mutation responsible for mature onset diabetes of the young. Starting as a bench scientist, I eventually became a vice president running nine different departments ranging from sequencing and molecular pathology to assay development—all applied technologies necessary for developing drugs. When I joined Millennium in 1993, the company had 10 employees; when I left it in 2006, there were 1,600.

For its first 10 years, Millennium licensed its technology to large pharmaceutical companies and set up collaborations with these companies to identify disease target genes so they could develop new drugs. Later on, Millennium had its own drug products, whose genetic targets we identified using the technologies and expertise we had developed.

One of my mentors at Millennium was Mark Levin, the company's CEO for 12 years. Mark was very passionate about the value of teams. He also made employees believe they could accomplish more than they ever thought possible. In fact, the motto by which he built the company was on our first T-shirt: "Nothing Is Impossible." Mark had us believing we could walk on water as long as we did it together.

I realized the importance of working in teams during my first project at Millennium. It was a very competitive project, and three of us were working very closely together. There was an excitement and thrill of working well toward an ambitious goal. Being able to complement and support each other and then in the end to beat the competition—that was great reinforcement of great teamwork.

FROM SCIENTIST TO MANAGER

At Millennium, I learned to appreciate teams and then found out that I enjoyed leading them. As I moved from bench scientist to manager, I needed to change how I thought about my role and contributions. I lost track of some of the technical details—once I was managing several hundred employees, I could no longer be a technical expert at what they all were doing.

Instead of being the technical expert, I needed to become comfortable enabling others to be successful and supporting them. One of my roles was to make sure employees were asking the right scientific and strategic questions. For example, I would question whether a particular experiment was really necessary and whether the results would help us make an important programmatic decision.

After I had been at Millennium for 11 years, I told Mark Levin that it was time for me to leave. I felt as if I had been a high school senior for too long and, having completed all my requirements, wanted to experience something new. Mark said to me that there would be a time to leave but suggested that I first learn about the corporate side of the business. I then moved to the corporate side of the company and was able to expand both my leadership skills and learn the business aspects of running a company. I stayed for two more years and learned a tremendous amount, but my goal was to go back to a start-up company. After leaving Millennium in 2006, I helped found Cerulean Pharma and was there for seven years. Then I helped launch Editas Medicine, a genome-editing company dedicated to treating patients with genetically defined diseases. Most recently, I have started Cedilla Therapeutics, a company focused on the field of protein degradation.

Since 2001, I have also been a director of Women in the Enterprise of Science and Technology, a Boston-based nonprofit that supports women in science and technology. We have monthly workshops in management and leaderships skills to help women be successful in their jobs. We believe that community is very important. We aim to create a safe community where women can excel and freely discuss career issues.

ABOUT THE SUBJECT

Alexandra (Sandra) Glucksmann is president and CEO of Cedilla Therapeutics of Cambridge, Massachusetts. She has also held executive positions at Editas Medicine, Cerulean Pharma, and Millennium Pharmaceuticals. Glucksmann is a molecular geneticist with a BS from the University of Wisconsin–Madison and a PhD from the University of Chicago.

2 A LAB OF MY OWN

Ethan Perlstein

For my career, I had planned to stay in academia and become a professor. That was my career goal because I emulated my adviser. He was the person I saw all the time; he was my role model. I identified with him and wanted to be like him. As a professor, I planned to work in the area of academic drug discovery, creating the germ of the idea and then remaining involved in companies that would be spun out of my lab. Although I enjoyed teaching very much, getting to teach wasn't my main reason for staying in academia. My primary motivation was to engage in research, and I believed the

way to do so was in an academic lab. That was all I knew. I didn't have a sophisticated understanding of the grant process, university funding, or the overall biomedical system. But after being turned down for 27 academic positions, I realized that becoming a professor wasn't going to be my career direction.

So I took the unusual step of becoming an independent scientist and opened up my own research lab to identify drugs to treat rare diseases. I initially raised money from angel investors. As a solo founder, I created what is now called Perlara in February 2014. Two years later we were negotiating financing from a biopharma partner with whom we planned to conduct a multidisease collaboration. Our current goal is to demonstrate that we have a viable business model and together with our partners to demonstrate that the science is useful. We need to find molecules to advance toward clinical trials, setting the stage for our first round of venture-capital funding. There are 4,000 rare single-gene diseases, and we need to figure out for which of them our approach and the economics make sense.

HIGH SCHOOL BIOTECH

I grew up in North Miami Beach, Florida. My first engagement with science was reading *Scientific American* when I was a teenager. I was fascinated by biology. My parents weren't scientists, but my mom was an unconditional supporter of my scientific interests, and my dad encouraged my general intellectual curiosity.

> Once I'm set on something, it is difficult to dissuade me from doing it, come hell or high water. That's been a part of my character resilience throughout life. My parents went through multiple divorces, and there was a lot of emotional difficulty in my childhood. So I learned a constitutional resilience from an early age. I see now how once I became interested in science, that resilience became a kind of single-mindedness, which has permeated how I approach science from day one.

My first lab experience was a summer internship at a biotech start-up after my junior year of high school. At first, I worked in a chemical stockroom mixing solutions and growth media. After the summer, I was promoted and got to work directly with two scientists on a phage-display research project using bacteriophage to express T-cell receptor proteins. I worked there every day after school during my senior year. I was thrilled to be working in a lab and looked up to the scientists as role models.

When I was working at the biotech start-up, I spent time reading immunology journals in its library. One of the papers was by Ronald Germain, who is a premier T-cell biologist and currently the chief of the Systems Biology Laboratory at the National Institute of Allergy and Infectious Diseases at the National Institutes of Health (NIH). I sent him an email with questions about his work. In fact, I sent emails to quite a few scientists, but Ron was kind enough to respond to a high school student. This response led to a correspondence and then to an internship during the summer before college. I returned to the NIH for the summers after my freshman and junior years and to the biotech start-up in Florida after my sophomore year.

While in high school, I often visited my grandfather, who was living in New York, and I fell in love with the city. As the great-grandchild of Polish

Jewish immigrants who had immigrated to the United States through Ellis Island and spent their lives in the city, I felt a cultural connection to New York and decided I wanted to move there. So in senior year of high school, I applied for early admission to Columbia University and was accepted.

As a freshman, I took chemistry, calculus, physics, and biology with the intention of becoming a biologist. But I didn't keep up with my classes and didn't take homework seriously. I hadn't made the transition from high school, where I could get by on little study, to the more serious mentality needed for college. My initial grades in science courses were not stellar, and I started to question whether I really wanted to be a scientist. I continued to work in biology labs each summer, but during the semester I enrolled mainly in humanities courses. I majored in history and sociology and considered for a while becoming a social scientist. I was interested in the period in which my great-grandfather had immigrated to the United States. I studied the Polish language and worked with a professor in an independent study about the Poles' society. My mother had spoken Polish to me when I was younger, so I already had a basic working knowledge of the language, which I solidified by taking language courses.

SOCIAL SCIENTIST TO BIOLOGIST

But by senior year I decided I really did want to be a biologist and took laboratory and research courses to make up for lost time. I applied to several graduate schools, but, as a humanities major, I was not sure if I would be accepted. I was admitted to several biology departments, though, including those at Berkeley and Harvard. I asked my NIH mentor, Ron Germain, which program would be best for me. He recommended Harvard as a place where I could grow either in immunology or in another area of biology, so I decided to study there.

I've met enough first-time founders to see that we share certain characteristics. Those characteristics are not as rare as a rare disease. Above all, it's a kind of persistence and being inured to failure. I'm passionate about what I am doing. I'm also aware of some privileges that I have that many others do not. I had enough financial resources to get started, and I was privileged to have spent time as a fellow at Princeton.

Harvard's system of helping graduate students find an adviser and an area of research includes having the students work for a short time to try out various labs. My first rotation was in a stem cell laboratory, but I didn't like cutting up mouse embryos—it made me squeamish. I decided I wanted to work with an organism in which I could get data more quickly than with mice, and I wanted an area where I didn't have to cut up animals.

For my second rotation, I worked with yeast cells in Stuart Schreiber's interdisciplinary lab combining chemistry and biology. Schreiber is a leader in chemical biology, which uses small molecules as probes in uncovering biological functions. I much preferred working with yeast cells to working with mice. For my PhD thesis, I worked on two projects: the first in the field of drug discovery, looking for candidate drugs using yeast cells and a simplified disease model, the second in the area of yeast genetic variation, either created artificially through mutagenesis or already existing naturally in wild yeast populations and that I could breed in the lab. My goal was to identify mutations in yeast that make them either more or less resistant to a drug. I made up the term *evolutionary pharmacology*, to describe the approach. It is still an open question and a key one for my company: To what extent can one use yeast to gain insights that will be useful for pharmacology in a clinical setting?

Stuart Schreiber had a large research group with several dozen postdocs and graduate students. It was a great environment in which to train, with lots of resources and mentorship provided by the more experienced students. I realized early on that I wouldn't get too much of Stuart's time and that his mentorship would be primarily as I was finishing the thesis. Every week, during lab meeting, he asked good questions. Then once my project matured, and I was writing up the results, Stuart became fully available, and we worked together more closely. I was in grad school for five years and then did a transitional postdoc year in Stuart's lab finishing experimental work and writing scientific papers before I got my next position, a fellowship at Princeton University's Lewis-Sigler Institute for Integrative Genomics.

In the field of chemical biology, a postdoc works in a faculty member's lab, whereas a fellow works on his or her own but reports to the head of a center. A fellow is at the institution for some time to prove himself or herself in a discipline distinct from his or her graduate studies. So a fellow is like a "super-postdoc." Fellows work independently but also need to obtain their own funding and are able to operate small groups on their own.

As a fellow at Princeton, I continued to use yeast as the experimental system, concentrating on the effects of a particular molecule, the antidepressant Zoloft. I took a number of different approaches to create a model of how Zoloft works in a yeast cell, including genetics, biochemistry, pharmacology, and radio labeling. I was able to make connections from yeast to mammalian cells, showing that similar interactions with Zoloft occur in both systems.

Persistence has led me to try to achieve my goals in whatever way I can, using whatever tools I had available at the time. I call that trying to "hack" the system. Twenty years ago, when one could first email a scientist, I used my father's email account because I didn't have my own email account. As a high school student, I was able to convince a well-known researcher at NIH to offer me a summer internship. I figured out a way to connect a scholarly journal library in Miami to researchers around the world. By creating my own research lab outside of the standard academic setting, I've managed to "hack" the system in order to achieve my career goal.

LEAVING ACADEMIA

I arrived at Princeton in 2007 expecting my fellowship to last a year or two. But when the recession occurred in 2008, colleges stopped hiring new professors. I thought the academic job market would get better in a few years. Indeed, by 2010 the hiring freezes had ended, but by then there was a large supply of postdoc candidates like me available for just a few positions. Over the next two years, I slowly came to the realization that I would have to give up on my fantasy of becoming a professor.

In 2012 and 2013, having left academia, I went through a period of learning, not knowing where my career would head. I decided to set up my own research lab. First, I found a space. I considered carrying on my previous work, but I concluded that would be too difficult to commercialize. As I spent time interacting with rare-disease advocates, I saw a commercial opportunity: the most promising application for the same evolutionary pharmacology I had been working on would be in the area of rare diseases.

The first lab space I leased was a little "nook" in downtown Berkeley. I was self-funding, paying for the research from savings. I started with some modest experiments using yeast. After six months, I was running out of money, so I found a job as a consultant for a start-up then called Microryza and now called Experiment, a kick-starter for science projects. Experiment was part of Y Combinator, Silicon Valley's premier start-up accelerator primarily known for backing companies such as Airbnb but at this time branching out into the biotech area. Using what I learned at Experiment, I incorporated Perlara (initially named Perlstein Lab) as a public-benefit corporation, a for-profit company that balances profit and mission in a way that would allow me to raise capital and hire employees.

During the first two years, Perlara was in the seed stage, building a business model and becoming scientifically differentiated from its competition. In our third year, we entered into collaboration with Novartis to develop therapeutics for a group of rare, inherited metabolic disorders. In 2017, we created a program in which we partner with patient families to jointly fund cures for their rare diseases. We begin by studying the disease in model organisms, progress to a drug-repurposing screen, and culminate the process with a drug-screening campaign to discover compounds suitable for preclinical validation in mice. The patients' families become join owners of the intellectual property, which could then lead to creation of a new company to manufacture the therapeutics.

Perlara is now in its fifth year, and we're well on our way to demonstrating a sustainable business model. I am happy to have founded a company in which I have the opportunity to do research on rare diseases, although not in the academic setting, as I had originally envisioned.

ABOUT THE SUBJECT

Ethan Perlstein is the founder and CEO of Perlara LLC of San Francisco, California. He has a degree in sociology from Columbia University and a PhD in molecular and cell biology from Harvard University.

3 A MATHEMATICAL MODELER

Beth Reid

I figured out what I wanted to do for a career during a summer internship held at NASA Langley before my senior year in high school. I was working with a chemical engineer doing combustion experiments using two counterdirected gas streams with a flame in the middle. A scientist walked into the room, looked at our experiment, and started asking questions. I asked who the researcher was and was told he was a mathematical modeler. At that moment, I decided I, too, wanted to be a mathematical modeler. It turned out to be the right assessment of my interests, and it became my career.

I did mathematical modeling in astrophysics as a graduate student at Princeton, analyzing data from the Sloan Digital Sky Survey and creating models for how galaxies are distributed among dark-matter halos. I continued this work as a postdoc in Barcelona and at Berkeley. I loved my time as a postdoc. It was glorious to have the freedom to do research all day, work independently, and develop new ideas.

But I also wrote my first grant proposal as a postdoc, and the process of writing it made me realize I didn't want to continue in academia. The proposal was to apply my expertise in the modeling of galaxy clustering to a new NASA mission called WFIRST. Even though the proposal was funded, by the end of writing it, I was already feeling tired of working in a single narrow area. I imagined having to spend my time as a professor writing proposals and teaching, with little time left over to work directly on my research. I envisioned dragging my husband from the Bay Area to a small midwestern college town just so he could watch me subject myself to the stress of the tenure process. Even though my mentor at Berkeley tried to convince me otherwise, I decided that academia was not the life for me. I began looking for an industrial job that would make use of my mathematical modeling experience, especially my experience analyzing large data sets.

I started my job search by talking to friends and former colleagues. Some of my colleagues were employed by companies doing analytics on web data. Such companies look at product usage and investigate how consumers behave. But I quickly learned that this area didn't interest me. Then I saw an ad on LinkedIn looking for a researcher to do spatial-temporal and Bayesian hierarchical modeling. The company advertising the position, the Climate Corporation, is a recently acquired division of Monsanto that offers precision agricultural products to farmers. I generally understood the mathematical terms in the ad, and the job sounded scientifically challenging. I figured I could fake my way through an interview.

With my background in astrophysics, getting hired by this company was not a straightforward process. Though my skills overlapped with many areas of research in the Climate Corporation, I didn't fit squarely into any single area. After I interviewed with several teams, the meteorology team said, "We like her!" I am now a data scientist on a small team that creates models using satellite, radar, and rain-gauge data. We provide farmers with recent precipitation information to aid in day-to-day decision making, and we feed this information into many other internal models, such as our soil-nitrogen model. Even though I'm just beginning to learn about weather and radar data, I am enjoying working in a new field with an environmental focus. My rigorous scientific and mathematical training has proved invaluable and quickly made me a productive member of the Climate Corporation meteorology team.

Compared to many women scientists, I have been less affected by social pressures, gender bias, and overt discrimination. I have tended either not to notice or to be unaware of situations affecting others. I learned years later about discrimination in graduate school and during my postdoc of which I was unaware at the time. Had I had been more attuned to [these situations], I might have been more helpful to my colleagues.

AN EARLY INTEREST IN MATH

I grew up in a suburb south of Richmond, Virginia. My father was a chemical engineer, and my mother was a librarian and taught elementary school. I attended a public high school that specialized in math and science. I was definitely interested in math. One of my favorite teachers at the school was the calculus teacher. She was a talented, supportive teacher, and I am still in touch with her. As a young person, she wanted to study engineering, but she believed that as a woman she did not have this option, so she became a math teacher.

During a junior-year class, each of us listed a favorite book. One of my classmates mentioned *A Brief History of Time* by Stephen Hawking, which I then read. I found the concepts fascinating, and after reading several other popular physics books, I decided I wanted to pursue physics. That summer I was one of 10 to 15 high school students doing research as part of a special "governor's school" program for Virginia residents at NASA Langley when I decided that the type of physics I wanted to do for a career involved mathematical modeling.

For college, I considered Carnegie Mellon but decided instead to attend Virginia Tech. The in-state tuition was much lower, and the chair of the Physics Department, Lay Nam Chang, sold me on the school when I visited. Virginia Tech is primarily an engineering school with a large physics faculty and only a few physics majors each year. This gave me plenty of opportunity to work on research projects and get individual attention from faculty members. It was a wonderful place for me.

At Virginia Tech, I worked on research projects with condensed-matter physicist Uwe Tauber for several years. I had a work-study job doing computational statistical mechanics. We were studying reaction-controlled diffusion, and the computational model predicted phase transitions as a function of various interaction parameters. I wrote code to identify the phase transitions and to test Tauber's theory. Tauber was a very hands-off adviser, and although we met regularly, he let me figure out the best way to write the code to test his models. I appreciated the independence, but it definitely resulted in unnecessary headaches and wasted time. For example, I was doing computation in Microsoft Excel, which does not scale very well, to visualize 10,000 particle locations. Over the years since then, I have taught myself good coding technique, but I wish that I had been given formal training in computation as an undergraduate.

I met my husband at Virginia Tech. He, too, was a physics major, although we didn't have many classes together. He stayed at Virginia Tech for three years after I left for graduate school. During that time, he studied for master's degrees in teaching and area studies, which is a blend of cultural theory and history focusing on a particular geographical region. Then he moved to New Jersey and taught high school physics for two years while I finished my PhD.

Instead of Princeton, I considered going to graduate school at the University of California–Santa Barbara to continue working on computational statistical mechanics. There was a group working on modeling of forest-fire propagation, and I have always had an interest in ecological and environmental topics. In fact, I wrote in my National Science Foundation Fellowship application that I was considering using physics models to study animal habitats. But I didn't want to commit to the forest-fire project and then have few other environmental topic choices for a thesis.

My decision to attend Princeton for graduate work was based more on a feeling than any careful planning. When I visited, one of the students pointed out the room in the graduate college where the renowned theoretical physicist Richard Feynman had lived when he was a student there. I am a Feynman fan, and as an undergraduate I had read some of his books. Seeing his room was what convinced me I should attend Princeton.

LOOKING FOR A FIELD AND A TOPIC

But once I arrived at Princeton, I had a difficult time finding an intellectual home. Although the professors were happy to chat with me, they expected students to be very independent. I looked around, but I couldn't find a good problem to work on. My first possible thesis topic was a modeling study in the Ecology Department. Next I did an experimental project with a biophysicist, who told me point blank that I was a theorist and not an experimentalist. Then I considered a project in neuroscience. After three experiences with not finding a research topic that was the right fit for me, I began to lose confidence that I would find my niche.

Finally, I considered astrophysics. David Spergel was very receptive and happy to talk to me. He started by giving me a very straightforward problem. When I started on his problem, I discovered that the very first equation in the very first published paper I read was incorrect. I often tell this story as a good life lesson. I spent days making myself absolutely sure that this first equation was wrong. But it did turn out to be the case—I was correct.

After completing my first astrophysics problem, I met with David once a week to chat. As an adviser, David encouraged independence and wanted me to come up with my own thesis topic. That was the same issue I had with the other professors—I preferred that they just give me a cool problem to work on. As a beginning graduate student, I didn't believe I had the depth of understanding or knowledge of the field to carve out a problem on my own.

For a semester, Dave and I had random conversations, and I did a lot of reading. I was feeling frustrated; my graduate school career was not going well. At that point, I met a postdoc working on measurements of the cosmic microwave background and asked for a project. I worked for a while on the pointing of the Atacama Cosmology Telescope (ACT), which is a large radio telescope in the Atacama Desert in northern Chile. The ACT had a 1,000-detector array, each of which needed to be recalibrated periodically to maintain its pointing accuracy. One of the collaborators on the project was a professor from another university. We had a meeting in which I presented my results to the group. He arrived just in time for my talk, stormed into the room, and declared, "I am not going to believe a single word that you say." It was quite intimidating. He was in charge of pointing the telescope, and the inaccuracy values that I was presenting were large. It turned out my values were correct, but he didn't want to hear that.

I have had other experiences in which professors have been aggressive and confrontational and have told students and colleagues that they were flat-out wrong. I have gotten used to this behavior, and now I just play along. But when I was a grad student, it was disturbing to me. My method of coping with such behavior is to be absolutely sure I am right. Since I am an extremely careful person, in such a situation I have never been wrong. I take comfort in that knowledge.

For my thesis, I used data from the Sloan Digital Sky Survey to estimate fundamental cosmological parameters, such as the total matter and size of the universe. This required my understanding complicated physics, such as how galaxies form and where they form in relation to the underlying nonlinear perturbations that we know how to calculate. At the time, there were disagreements about the cosmological parameters to be inferred from the data. One researcher was measuring and modeling the galaxy power spectrum in one way, and another researcher had a different method and was getting a different result. David suggested that I should try to figure out which approach was correct, and I did. David was quite busy while I was working on my thesis—he was chair of the Astrophysics Department and supervising many students. At the time, I found it stressful to be working with limited guidance, but over time his approach inspired me to be independent.

At the end of my time in graduate school, I got lucky because the final data from the Sloan Survey was just being released. Most of the researchers working on the survey were moving on to other projects. Somehow it fell in my lap to write the final summary paper of the galaxy power spectrum as measured by the survey. So I got to be first author on an important, highly cited paper. This has been my crowning scientific achievement. It happened not only because I was qualified and knew the subject well but also because I was in the right place at the right time.

After graduate school, I went to Barcelona for a postdoc to analyze the final Sloan data and write the summary paper. That took a year. I worked on various projects the second year and then moved to Berkeley as a Hubble Fellow to work on the next generation of Sloan Survey data. By then, I was an expert in this field, coming up with my own ideas and learning how to execute them. It was a great time scientifically, but it also meant that I was continuing to work in a narrow research area, and I was itching to try something else.

As I looked around to decide the next phase in my career, I realized that moving from being a postdoc to being a professor is a significant career change. Writing grants, teaching, mentoring graduate students were all activities that I didn't have to do as a postdoc. I loved doing research, but the other job responsibilities of a professor were not appealing. So I decided to leave academia. I'm pleased to have found an industry job in an area related to my longtime interest in the environment that also makes use of my talents as a mathematical modeler.

ABOUT THE SUBJECT

Beth Reid was a scientist on the weather science team at the Climate Corporation. Since the printing of this book, she is a software engineer for balloon planning and control at Loon LLC. She has an undergraduate degree in physics from Virginia Tech and a PhD from Princeton University.

4 A LONG-STANDING INTEREST

Christopher Loose

I have had a long-standing interest in creating new pharmaceuticals. Even in middle school, I was interested in medical therapies, thinking of either becoming a medical doctor or working in medical science. As I got older, I realized I could help more people by doing drug development than I could by helping one patient at a time. I decided to study chemical engineering with a focus on biomedical applications as well as learning about business. My interest in both pharmaceuticals and business led me to start two companies: one during graduate school and a second one just recently.

My first start-up was a medical device company called Semprus BioSciences. We developed vascular catheters with a surface modification designed to help reduce clot formation. From that experience, I learned the power of partnering and teamwork and how long it takes to grow a company from idea stage through development and finally acquisition.

After David Lucchino and I cofounded Semprus, we examined a range of technologies on which a new biomedical company might be based. We considered ideas, for example, in regenerative medicine using genome editing. But that technology is quite complicated, with delivery challenges and possible adverse changes to the body's genetics. My MIT thesis adviser Robert Langer and his colleague Jeff Karp have developed a simpler approach with great potential. They have shown that small-molecule drugs can activate stem cells lying dormant within a patient's own body, which can grow and restore healthy tissue and function. David Lucchino and I decided to create a company based on Langer and Karp's research to bring novel small-molecule regenerative therapies to market.

Our new company is named Frequency Therapeutics, and our first target is reversing hearing loss. There are millions of people with hearing loss, and many of those people have damage to the hair cells of their inner ear. Our drugs target stem cell descendants called progenitor cells, causing them to multiply and create new hair cells, potentially restoring natural hearing. We are well along in our preclinical work toward that goal and hope to be in the clinic in the not too distant future.

STARTING OUT

I grew up in Okemos, Michigan, a suburb near East Lansing, where my father is a lawyer and my mother was a teacher. I was interested in math and science from an early age and while in high school had the opportunity to do lab research at Michigan State University, confirming my interest in pursuing engineering. I looked for a college that was strong in those fields, offered good history and liberal arts courses, and was not too large. I was accepted at Princeton, my first-choice school.

At Princeton, I majored in chemical engineering. I also took business courses, including accounting, risk management, and probability, and satisfied the requirements for a certificate in operations research. During senior year, I worked with Professor Chris Floudas on a research project to create an energy model for protein folding. My senior thesis became part of a computer model for therapeutic peptide design that Floudas's group used in its collaborations with pharmaceutical companies. I found it very exciting when our quantitative work was directly applied to solving medical problems.

Professor Floudas was an important mentor for me while I was at Princeton. He impressed me with the clarity with which he approached problems and his ability to analyze complex systems, from protein folding to chemical-plant manufacturing. He taught me to make progress on large, messy problems by breaking them into smaller problems, coming up with appropriate assumptions, obtaining approximate solutions, and then using those solutions to gain insight into the original systems.

AN INDUSTRIAL EXPERIENCE

After Princeton, I joined a development group at Merck & Co. in nearby Rahway, New Jersey. My time there was a valuable experience. My colleagues at Merck really believed in the importance of new medicines, were dedicated to their jobs, and were very good scientists. I learned how pharmaceuticals are scaled-up to use in clinical trials and how companies transition a drug into pilot-plant production. Surprises often occur when scaling-up drug manufacturing, and I saw some of them firsthand. I also became familiar with the documentation that accompanies production as well as the environmental and safety concerns.

When I joined Merck, I wasn't sure what the best next step would be for my career. Although I was thinking about going to graduate school in chemical engineering or possibly getting a business degree, I decided to find out what industry is like before making a commitment to pursue a PhD or study for an MBA. Working at Merck convinced me that creating pharmaceuticals was what I wanted to do—it was technically and professionally engaging to develop drugs. But I also saw how hard it would be to have much impact in such a large organization. And to make a difference in such an environment, I saw that I would need an advanced degree.

One of my close friends from Princeton had gone directly from college into a chemical engineering graduate program at MIT. When I visited him to learn more about the department, I was impressed with the students' and faculty's energy and enthusiasm and the opportunities for innovation and entrepreneurship. The visit convinced me that I should pursue a PhD rather than an MBA and that I should start the program as soon as possible. Shortly thereafter I applied to MIT and so spent only one year at Merck.

I entered MIT committed to finding a thesis project with commercial potential and to working with Professor Bob Langer if possible. Langer is a world expert in drug delivery, tissue engineering, and biomaterials. He is very supportive of students and colleagues as they work on difficult fundamental and applied problems. Langer has been involved in many start-up companies, and I knew that joining his lab would give me access to a network of colleagues in an environment supportive of innovation and commercialization.

I was fortunate to win a Hertz Fellowship, which allowed me to work on an unfunded project with Langer and Professor Greg Stephanopoulos designing antibiotics using a computational algorithm. When the new antibiotics proved effective against Methicillin-resistant *Staphylococcus aureus* and anthrax, I approached Bob to discuss a path to commercialize the new drugs. One choice was to start a drug-discovery company. But after learning about regulatory and market hurdles, we realized that a better, shorter-term commercial opportunity was in the area of antimicrobial medical devices such as oncology catheters, which are frequently a haven for infection. Bob was supportive of the idea and introduced me to David Lucchino, a Sloan Fellow at MIT doing a one-year, midcareer MBA.

David turned out to be an ideal business partner with a complementary background and skill set. He had worked on Madison Avenue in technology marketing and had been a leader of a biomedical seed fund before coming to Sloan, whereas my experience was more directly technical. What we shared was a competitive drive and the desire to make a big contribution in the biomedical field.

David and I spent time at Beth Israel Hospital, where we became friends with a nurse who had spent many years treating catheter-related infections. We talked to infectious-disease doctors and to hospital administrators involved in purchasing and value analysis. Based on many conversations with these stakeholders, we decided our product had significant commercial potential. The energy and excitement that I felt thinking about the next steps to creating our product was a sign that entrepreneurship was the right path for me.

ENVISIONING A COMPANY

Although David and I wanted to create a medical device company, the first challenge we faced was that neither of us had touched, let alone worked with, a medical device in our lives. We needed help from someone with experience building and selling catheters. On a whim, David dropped by a medical device regulatory conference and learned that a leading antimicrobial catheter company was downsizing and laying off experienced staff. David

The SteriCoat team consisting of Chris Loose (*second from left*), Joel Moxley, David Lucchino, Mike Hencke, and Vipin Gupta (*not shown*) won the MIT $100K Entrepreneurship Competition in 2006.

met Greg Haas, who had been developing catheter products for more than 20 years. Greg taught us about device manufacturing, the regulatory process, and the questions we should expect from potential commercial partners.

David and I wrote a preliminary business plan and decided to enter it into the MIT $100K Entrepreneurship Competition. Win or lose, we knew that competing would help us make additional connections and improve our plan. The spring of 2006 was very busy as I continued research in the lab, started to write my dissertation, and prepared for the competition. After getting feedback from dozens of great advisers, we won the top prize in the competition. This gave us $30,000 and some confidence that our start-up, SteriCoat (later renamed Semprus BioSciences), was heading in the right direction. One adviser joked that our early business model was to apply to every business-plan competition in the world with open admission: we had also entered competitions at Harvard, Oxford, Princeton, Cambridge, and Rice Universities, accumulating about $100,000 in funding.

The next step for our venture was pursuing angel funding from affluent individuals. Recognizing that taking private investments is a very large responsibility, we held off for six months while we made sure that we had sufficient intellectual-property protection and recruited the rest of our team. I also worked overtime to finish my thesis in order to graduate in the spring of 2007.

A big break with private investors came at a meeting with Harvard Business School professors who sometimes support early-stage companies. As we were halfway through our first pitch, one professor leaned back and said, "I'll put in $50,000." Without missing a beat, David audaciously replied that we were actually looking for $100,000 checks. As I watched in amazement, the professor quickly agreed. We then met with the professor's friends and a few others and raised $1 million in start-up funds within a couple of months. After raising private funds, we met with venture-capitalist firms, quickly proceeded through due diligence and added an additional $1 million in July. How good times were in 2007...

BUILDING A COMPANY

Now the building began in earnest. The work was intense, deadlines were brutal, and the effort turned into a grind within the first year. We tried to grow the company at a disciplined pace and not hire too quickly. By the end of 2008, we had about 10 employees, and I was CTO. A larger team meant that more time was devoted to meetings and management. There was also the constant pressure to raise additional funds.

Our challenges really began with the market crash in 2008. After getting a long string of "no's" to our funding requests from venture-capital firms, the leadership of our primary backer, 5AM Ventures, proved crucial. 5AM Ventures told us it was committed to funding our next round regardless of other investor decisions. We were able to find a second investor just in time and added SROne, associated with GSK, in a later round. The key thing we learned at this point was that having committed, experienced investors with deep pockets is crucial for a start-up to survive difficult times.

Our next challenge was receiving information that the market was shifting and that we needed to focus on blood clotting more than on infection prevention. We also heard that the Federal Drug Administration (FDA) might increase the hurdles to review new antimicrobial technologies. For both reasons, we shifted from the purely antimicrobial technology that I created at MIT to a dual-function technology that we licensed. Even though I was sorry our core technology was no longer my invention, it was much more important to focus on market need than on pride of inventorship.

The work remained challenging and intense through 2012 as we got closer to FDA submission. Our company had grown to 40 employees, and my responsibilities were wide ranging and seemed to change every six months. I worked on creating intellectual property, performing regulatory analysis, planning clinical trials, analyzing experiments, and diversifying our pipeline with multiple projects. We recruited mid- and late-career experts, from whom I was constantly learning new areas. The most important role that David and I had during this phase was to recruit the best team members and rally them to answer the key questions driving us to meet our milestones.

The year 2012 was pivotal. We submitted to the FDA an oncology catheter that would reduce clotting; we had a contact lens designed for improved comfort in early human trials; and we were beginning to get acquisition interest. It had been five years since our early investors provided financing. The board ultimately agreed that our best option was to sell the company to Teleflex Medical. After the acquisition, David and I remained at Teleflex to witness FDA approval and to oversee the transition, but neither of us was interested in becoming long-term employees in a large corporate setting.

EXPLORING NEW OPPORTUNITIES

It had been eight years since I started working on medical device surface modification: 18 months at MIT, five years at Semprus, and 18 months

Chris Loose (*center left*), his longtime partner David Lucchino (*center right*), and members of the management team of Frequency Therapeutics gather outside their start-up headquarters in Woburn, Massachusetts, in 2017. In addition to Loose and Lucchino, also shown are Rick Strong, Jack Herman, Raj Manchanda, and Mike Jirousek.

at Teleflex. That felt like a long time to work in a narrow technical area. I decided to explore other medical ideas, with the possibility of cofounding another company with David. I also have always loved teaching and mentoring students and looked for ways to combine these interests. One of the ways was by joining the Center for Integration of Medicine and Innovative Technology (CIMIT) as a business start-up accelerator executive.

CIMIT is a health-care consortium that links research centers, universities, and medical device affiliate companies in the greater Boston area. The consortium receives proposals for dozens of new company concepts each year and provides mentors and funding. While at CIMIT, I also assisted new Langer lab start-ups with science and intellectual-property strategy and felt energized by my connection to students and faculty at these early-stage companies.

I also heard about an opportunity to work directly with students in an academic setting. A friend at the Yale Medical School told me about a new center that was being established to commercialize university-developed biomedical technologies. I became director of the new Center for Biomedical Innovation and Technology as a part-time adjunct professor, helping students and faculty evaluate a range of biomedical ideas, including health information technology, drug delivery, medical devices, and diagnostics. I also developed and teach a health-care ventures course.

During the year after David and I left Teleflex, we actively researched ideas for our next company. We probably discussed a hundred possibilities and did a careful study of about a half-dozen. The most promising idea came out of a discussion with Professor Jeff Karp, a researcher with a joint position in Bob Langer's lab at MIT who has his own lab at the Harvard Medical School. Karp and Langer had been doing research on LGR5 intestinal stem cells. These cells are remarkable because they are able to replace the intestinal lining every four or five days, creating the full range of cell types present in the epithelium. Karp, Langer, and their groups made a breakthrough discovery: they figured out the pathways that control those cells and identified small molecules that can either inhibit or activate the pathways. Langer and Karp asked whether those same small molecules could be used to activate stem cells that lie dormant in a patient's body, a process that would have advantages compared to delivering stem cells to the body from the outside. Their idea is to give the progenitor cells a temporary "kick" with a molecule, causing the cells to grow and potentially restore healthy tissue and function. This is a new mode of therapy with a great deal of potential that has many possible medical applications. The first application we decided to work on is restoring hearing, and our first target organ is the ear.

There is an enormous unmet medical need for a therapy to restore hearing; there are millions of people with hearing loss whose only current option is a medical device such as a hearing aid or implant. The ear contains progenitor-like cells that our therapy can target. And because the ear is an isolated organ, you can do targeted drug delivery in the ear, with limited exposure of our therapy to other organs.

The cells in the ear that we will target are the LGR5 progenitor cells in the cochlea; these inner-ear cells lie dormant in mammals. We have proof of concept using animal studies showing that the same small-molecule drugs that control the LGR5 cells in the intestine can give a "kick" to the progenitor cells, creating new cochlea hair cells and potentially restoring natural hearing. Based on these studies and the large market potential, we have raised private funding and started a company called Frequency Therapeutics, headquartered in Cambridge, Massachusetts. If all goes well, our therapy will be in clinics to benefit patients with hearing loss in the not too distant future.

ABOUT THE SUBJECT

Christopher Loose is the co-founder and Chief Scientific Officer of Frequency Therapeutics of Cambridge, Massachusetts. He earned a BSE in chemical engineering from Princeton and a PhD in chemical engineering from MIT, where he was a Hertz Fellow.

5 START-UP OR LARGE VENTURE

Z. Jane Wang

At the end of my postdoc, I had the choice of continuing with a promising start-up in which I was involved or joining a new venture at Google. The start-up was a university spinoff using synthetic biology and protein-engineering technology that I helped develop. The Google venture was very secretive, but it was Google's first foray into life sciences and promised to be different from any biotech I previously knew. The advantage of a venture backed by a large company is that it gives you the ability to think about big problems and not be pushed to achieve financial results in a short timeframe Tackling longer-term objectives are often difficult in a start-up environment where monetary contraints force you to focus on tasks with the highest financial return.

I weighed my options and decided to join Google Life Sciences, now Verily Life Sciences. Even though my friends at the start-up have been quite successful, I believe I made the right decision. Rather than continue in my postdoc specialty, I learned to apply what I knew to new areas such as nanotechnology. Our team has a challenging goal—to improve drug delivery to organs or specific cells. Verily is developing new diagnostics to track the progression of illnesses, combined with more traditional Google approaches such as software to analyze large data and statistics to create a comprehensive health-care platform. At Verily, I am working with some great people and making a significant contribution to the team using my background in organic chemistry.

IMMIGRANTS AND PARACHUTES

I grew up in Honolulu, Hawaii, and attended public schools through middle school. I was interested in math and science, but the public schools in Hawaii were not strong in these areas. My father was a PhD student in meteorology at the University of Hawaii, and it was typical during that time for immigrant graduate students to bring their children to the university while they worked long into the night. The other kids and I would run in the corridors for hours on end. I remember tossing parachutes out the fifth-floor window to see how long they would take to reach the ground. We were encouraged to experiment and given the freedom to be creative.

My parents were from Beijing and came to the United States in 1990 with the first group of students the Chinese government allowed to leave the country to pursue graduate studies. These students were the cream of the crop in math and science in China. Because of the Cultural Revolution, many, like my parents, were older than most US graduate students. They were born in the 1950s and 1960s and were in their thirties by the time they became graduate students. The language barrier was an issue for them, so their technical skills were the ticket to their American dream.

As a child growing up in Honolulu, Hawaii, Z. Jane Wang was encouraged to experiment and given the freedom to be creative.

The combination of my parents' influence and my personal interest in building things led me to math and science. When I was midway through high school, we moved to Silicon Valley, where I found classmates who were much more academically inclined and motivated than my classmates in Hawaii. I had a very well-balanced high school experience: I swam and played water polo, enjoyed making pottery, and really liked literature. I graduated at the top of my class and decided to attend the California Institute of Technology.

Caltech is charming and quirky. The people are very enthusiastic, and that attitude is infectious. All the students are math or science majors, and everyone has to deal with the same courses and course load. That creates a bunker mentality in the students—so much pressure is a great bonding experience. When I arrived in 2003, my class was 25 percent women. Although we were a minority, we all were working hard in classes together, so I never really noticed that I lacked women friends. The experience seemed very inclusive as we struggled together.

I served on the admissions committee when I was an upperclassman, where we tried to be as gender blind as possible. At that time, the gender ratio at Caltech reflected the gender ratio of applications, which was predominantly male. Why that was true is complicated; part of it may be that by the time women graduate from high school, many are no longer interested in math and science. Women are often not rewarded for doing engineering and science, or, more accurately, they are rewarded for other traits. Things that dissuade young women from science can be as subtle as praising their traditionally feminine qualities rather than their acumen in engineering. I was lucky in that my parents were an exception; they encouraged me to become a scientist and praised me for being on the math team.

ENGINEERING MOLECULES

While at Caltech I gravitated toward organic chemistry. The concept of engineering molecules is tremendously powerful to me. Organic chemistry is the ability to make things; if electrical engineers make circuits, then organic chemists are the engineers of chemicals. The ability to dream up a chemical that might be useful for a specific application and build it in the lab is incredibly powerful. I experienced this firsthand while working with Dr. Robert Grubbs, who won the Nobel Prize in 2005 for olefin metathesis. Major League bats are made of Northern White Ash wood, which is increasingly rare and expensive, and the bats can be used only for a number of swings before shattering. Flying pieces of bat are very dangerous! Bob's lab took coarse wood and created a synthetic polymer that would fill the inside of the wood grain. The polymer bat is virtually indestructible and has the same mechanical properties as a fine wood bat. Unfortunately, professional baseball leagues have not sanctioned the use of Bob's bat, and it has now been discontinued, but the work allowed me to appreciate how one engineers molecules to achieve a desired physical result.

For my PhD work. I went to Berkeley to study with Dean Toste, a former student of Professor Grubb. I started with rather traditional organic-synthesis studies but transitioned to a field with more real-life applications. I focused on catalysis of coinage metals such as gold and silver. We built cages to encapsulate our metal catalysts and found that the resulting complexes behaved like the active site of an enzyme.

At Berkeley, I was able to surround myself with great people and had three thesis advisers. They were from different backgrounds, which was helpful because my project was quite atypical. Dean Toste was an early-career organic chemist, very sharp and quick. Bob Bergman, a pioneer in organometallic chemistry and physical organic chemistry, was thorough and encyclopedic: he listed things in steps: if this, then that, and so on. Ken Raymond, who worked on everything from magnetic-resonance imaging contrast agents to self-assembled nanocages, was my third adviser. With this diverse trio, I began a project using self-assembly to build cages with metals for catalysis: 3D enzyme-binding, sitelike structures from flat molecular building blocks. The project has now grown to six to eight graduate students and is widely published. It has been a very prolific joint project for all three labs.

The decision on what to do after graduate school was tough. Organic chemistry is a more mature field, so I wanted to try something new and different. I had tried to build an enzymelike catalyst from organic building blocks in grad school, so I decided I wanted to try to take an existing enzyme and get it to do new chemistry—to create for biology what I already knew how to do in organic chemistry. I chose to return to Caltech to work as a postdoc with Frances Arnold. Frances is a powerhouse of a woman. She pioneered the field of directed evolution of enzymes. I definitely look up to her as a mentor and a woman in science; she overcame a significant amount of gender bias in her early days as a chemical engineer to achieve success. Her style is to bring together people with diverse backgrounds. Her research group included a botanist, a chemist, an enzyme engineer, and even someone who had worked on zebra fish genetics. We had great resources in her group, but it was up to us to determine how best to collaborate and make use of them.

Z. Jane Wang and Nicole Peck won the DOW Sustainability Student Innovation Challenge Award in 2013 for creating novel enzymes using protein
engineering and synthetic chemistry. Wang and a group of her colleagues at
the California Institute of Technology created a start-up company to commercialize the technology. Also shown are Caltech physics professor Harry
Atwater and Resnick Institute director Neil Fromer.

CREATING A START-UP

A group of colleagues and I created a start-up based on our work in the
Caltech lab. There were definite commercial applications for the enzymes
we created. We met with pharmaceutical companies to show how our
enzymes could generate a faster and less-expensive production process
while also increasing yield. We could catalyze with iron what previously
had required expensive metals such as rhodium. Unfortunately, improving the production method of pharmaceuticals was not as profitable as our
initial estimates because chemical companies have a large investment in
existing infrastructure. We soon realized that the profit margin from material sales was not going to be sufficient to make business sense. Technically,
we had a great idea, but pharmaceuticals manufacturing was the wrong

application. This was my first foray into the business side of chemistry, and it was a great experience, but ultimately I decided to move in a new direction. Since then, this start-up has pivoted its platform, received venture-capital funding, and is doing well. The technology that I developed is still part of its core intellectual property, and I am glad my former colleagues found the right market for our technology.

Around the time that we were determining our market and business prospects, the Hertz Foundation organized a trip to Google. I was intrigued to see what Google was working on, so I joined the tour and submitted a résumé. I didn't expect Google to be interested in an organic chemist, but a few months later I got a call asking if I wanted to work on the secret Google X project. They couldn't give me any details—not even the area of the work. I assumed it was something relevant to my experience but nothing more. Even today my boss and I laugh about that. The project is one of the oldest ones in Google X and is now public. In 2016, those of us working on this project spun out from Google X and formed our own company, Verily Life Sciences. I have been working on functionalizing nanoparticles so they can target specific cell types. This would be very useful for both diagnostics and therapeutics but is a problem of enormous biological complexity. It has been fascinating to apply my background in chemistry and biology to these problems. My team has a mix people from both experimental and computational backgrounds, and I am learning something new from someone every week. The longer I stay out of academia, the less likely it is that I will return, but one never knows. I'm very happy with what I am working on at the moment. If in 10 years something I have worked on can help or improve a person's health outcome, I will be thrilled!

ABOUT THE SUBJECT

Z. Jane Wang is a technical team leader at Verily Life Sciences. She has a BS in chemistry from the California Institute of Technology and a PhD in organic chemistry from the University of California–Berkeley, where she was a Hertz Fellow.

6 A COMPANY WITH VALUES

Richard Lethin

In four years at a start-up called Multiflow, I learned a great deal about computers. We were struggling to make our circuits work through an onslaught of bugs in an era when one didn't simulate a computer before building it. The stress at work continued to build as we introduced the product into a competitive market. Sales were only slowly increasing as we battled to make our machine work despite faulty components. I was working seven days a week, often through the night, and hadn't taken a proper vacation in four years. We were discovering that the medium-scale integration logic components used in our designs had reached a fundamental limit. And so had I. My brain had rebelled, my eyes had stopped working, and I no longer could see the computer screen. It was time to rethink my direction and my priorities.

A meaningful life is about creation. The philosopher Friedrich Nietzsche wrote that the greatest act of human creativity is to mimic the gods by dreaming of new worlds and creating a new religion. Another great act of creation is to start a company with the right values. I'm pleased that over the past two decades I have built a company whose culture is based on ethical values such as fairness and honesty. Our consulting company, Reservoir Labs, is an independent small research laboratory that develops computer systems and solves problems in computer science, applied mathematics, and applied physics.

SCIENCE FAIRS AND NEW MATH

I was born in 1962 and grew up in Babylon, New York, a small suburban town on the south shore of Long Island. Some of the residents would take the commuter train to work in New York City. Others worked in the local high-tech and aerospace industries. Fairchild and Grumman had nearby manufacturing facilities. My father worked as a radar engineer for one such company, the Airborne Instruments Laboratory (AIL). My mother worked as a teacher at a nursery school. Both of my parents were supportive of my interest in math and science but let me find my own way and never pushed me in a particular direction.

I attended the local public schools, and one of them had some remarkable teachers. One outstanding teacher was Albert Kalfus, the head of the Math Department at the combined junior–senior high school. Kalfus had strong math, research, and education backgrounds and knew how to motivate and inspire a surly group of students. He was very ambitious on our behalf; to give us opportunities to get outside the school and compete, he created the Long Island Math Fair, which today is named after him. He also ensured that our school had access to a dial-up, time-shared, mainframe computer and lots of time and freedom to use it. He arranged for the seniors to teach seventh graders about the computer.

My seventh-grade math fair project was to figure out how to write the artificial-intelligence (AI) game *Animal*. It is a game in which a person starts by choosing an animal. The computer asks a series of yes–no questions, such as "Does your animal have four legs, or does it have a nose?" The computer then guesses the animal. If it guesses wrong, it asks for the name of the animal and a new question to learn for next time. I wrote the program in BASIC and had to figure out ways of storing the data, trees, file input/output, and the general algorithm.

The program got pretty "smart," and the game was a big hit at the math fair in 1974. We rolled the teletype into the cafeteria, dialed up the PDP-10 mainframe on our 110 Baud modem, and had the parents and other students sit down and "talk" to the computer. Everyone loved it.

By the time I was in seventh grade, the idea of communicating with a computer had been floating around in my brain for some years. When I was six years old in 1968, my dad took me to see the movie *2001: A Space Odyssey*. I was taken with the images of astronauts, spaceships, and moon landings, but what particularly caught my eye was the computer HAL, who interacts as a full member of the crew. The glowing ending of the film, when HAL is disabled with crystal blocks of its logic and memory floating inside its brain, has always stayed with me.

In seventh grade, Mr. Kalfus's ambitions and confidence in us meant we were taught New Math. It included abstract algebra: modular arithmetic, groups, fields, and rings. We learned about matrix algebra as a noncommutative ring and the Chinese Remainder Theorem. Many students and parents objected to New Math, and by eighth grade our class returned to a more standard math curriculum. But I loved the abstract mathematical thinking of New Math and later applied such concepts to my own work.

Through high school, I was one of a corps of three or four friends who spent lots of time dialing up the PDP-10, programming games such as multiuser *Star Trek*, as well as learning about operating systems and instruction sets. I also made some very good pocket money working for a local entrepreneur, Peter Locascio, who had spotted and recruited me at the math fair. He was buying the new personal computers, which I could only read about in *Byte Magazine*. This was prior to the release of the IBM PC; Peter was getting computers such as the Altair 880, which he wanted to sell to local businesses and doctors. I wrote the custom software for him to complete the solutions for those businesses. It was intense—working side by side with Peter in those offices outside their businesses hours to fix bugs, add features, and meet the deadlines that Peter had promised them. The solutions we built were actually pretty good: faster, more reliable, much less expensive than business computers sold by big companies and fully customized for each customer.

I decided that I wanted to go to college at Yale. It was mainly an emotional decision; I didn't put a lot of thought into it. My dad had been an engineering student there, and I decided that was also where I wanted to go. I majored in electrical engineering and took physics, math, and computer science. Yale has distribution requirements, so I also took courses in philosophy, political science, psychology, and art. My dad told me his favorite course was not one in engineering or science but one in French poetry. As I look back now, I also very much value the nonengineering courses I took and recognize how they shaped and enriched me.

But one engineering course was distinctly important. Junior year, I took a computer architecture course taught jointly by Josh Fisher and John O'Donnell. It was my first course covering current research topics and also my first engineering course not taught from a textbook. Fisher and his research group were advocating a new kind of architecture called Very Long Instruction Word (VLIW), which executed operations in parallel but used

After finishing an undergraduate degree in electrical engineering at Yale in 1985, Richard Lethin joined a start-up company named Multiflow to work on new computer architectures. After working there for several years, he decided to rethink his direction and priorities.

a compiler to schedule the parallelism. They compared their new architecture with other great architectures, starting with John von Neumann's original concepts from 1945 and later designs, such as the IBM 360, the Xerox Dorado, the CDC-6600, and the FPS-164, as well as Seymour Cray's high-performance vector machine. MIT professor Arvind visited Yale to talk about data-flow architectures, and Josh debated him vigorously about why VLIW was better.

ENGINEERING A START-UP

Josh Fisher was a visionary, and his enthusiasm was contagious. He was a good scholar, a good communicator, and a good salesman. Fisher had a vision for how a computer should be architected as a combination of hardware, circuit organization, compilers, and implementation. He sold me on his vision, and so, following graduation from Yale, I joined his and John O'Donnell's start-up, Multiflow.

With Josh's talent in selling, he was mostly away traveling, so John ran the show at home and was in charge of engineering. His style pushed the limits of his engineers and drove us nuts. He didn't see any other way than shooting for the stars, with reality as a secondary consideration. He wasn't mean or completely unempathetic about the impact of company demands on his employees. Rather, he saw possibilities. As a young engineer, I threw myself into the company and worked to my limit. If I had been more mature, I might not have been so influenced by Josh and John but able to judge their ideas and filter their requests. I might also have seen that the company was not headed toward success. But I didn't have the context or experience and so worked on the problems put in front of me instead of thinking about the bigger picture.

During my time at Multiflow, I observed my dad's career end at AIL. He had dedicated himself to the company and worked there for four decades. He had installed and serviced radars in the Berlin Airlift and then later at airports around the world and on US Navy aircraft carriers. The radar displays he designed resulted in large sales for AIL for decades. He rose through the ranks to become the director of engineering for aircraft landing systems. However, despite his loyalty and success, a series of acquisitions had distanced Dad from the "bean counters" who now controlled the company, and he was considered a cost. A few years short of being able to receive his pension, layoffs were threatened. Dad agreed to receive a lowball lump-sum payout and resign. He still had kids to put through college, but he never again was hired to work as an engineer. This experience instilled in me a deep distrust and skepticism about working for a "big company."

During my father's forced retirement, I noticed that his manager, who had a PhD, fared far better than Dad did during the reorganization. And at Multiflow where I worked, it was the PhDs who were in charge. After four years at Multiflow, I decided it was time for me to move on. I applied to graduate school and was accepted at MIT with the goal of obtaining a PhD and learning how to run my own business.

LEARNING RESEARCH AND BUSINESS

I entered the MIT Electric Engineering and Computer Science Department in the fall of 1989. My interest in AI, instilled at a young age by movies and science fiction, continued to guide my direction. I found a professor in the AI lab, Bill Dally, whose group was building a computer designed specifically for AI. After my intense training at Multiflow, I understood how to build computers. I dived in and helped Bill build a machine that worked.

I enjoyed graduate school. I enjoyed learning new areas, being part of the creative environment of the MIT AI lab, and having time for extracurricular activities and a social life. I played trombone in a local orchestra and made progress on engineering and controlling the AI computer. I was happy working on equipment, but eventually it was time to write a thesis. That was hard because it was the first time I was required to do research in a rigorous fashion.

My research involved understanding the architecture of the AI computer that we had built. The system worked well when lightly loaded but became congested and inefficient when operating near capacity, and no one knew why. That really bothered me, so I decided to study it as my thesis topic. I found that a fine-grained system with a high degree of parallelism needs special regulation. Studying that problem required me to learn math that I hadn't known, including combinatorics, probability, and queuing theory.

One of the members of my thesis committee, Tom Leighton, is one of the best mathematicians in the field of computer architecture. He forced me to be more rigorous in what I could claim to know; he forced me to do my homework. I say jokingly that Tom Leighton added at least a year to my time in graduate school as I learned and applied queuing theory to my problem and to his standards. But I value that year tremendously because the formal learning with Tom has paid off during my career.

To prepare myself to run a business, I called up a friend who had recently graduated from the MIT Sloan School to find out which courses were essential. He recommended product design and marketing, but the most important course he suggested was "Negotiation and Conflict Management," taught by Mary Rowe. Mary is a conflict-resolution specialist who spent more than 40 years as the MIT ombudsperson helping hundreds of people with serious concerns. The concepts from her course have served me incredibly well, especially techniques for integrative (win–win) negotiations.

When it was time to graduate, I had the opportunity to acquire a small consulting firm from friends who were moving on for other opportunities. I was able to negotiate a very fair "earn-out" agreement with them and good deals with new employees and so found myself running a profitable firm! The company is now called Reservoir Labs, Inc., and is headquartered in New York. We have about 30 employees, most of whom have PhDs.

We have research projects in cybersecurity; low-power computing for intelligence, surveillance, reconnaissance; compilers for exascale computers; compressive sensing, automatic reasoning; and big data. We also have clients for consulting projects in which we are developing compiler technology and algorithms for new kinds of computer architectures. We are shipping appliances for network computing for cybersecurity applications and ramping up with customers and prospects in major international companies and the federal government.

Looking back, I am pleased with the career path I chose. In contrast to the company where my dad worked, I have created a company based on values of honesty and ethical practices. As the sole owner, I can lead us into whatever technical investigation or business area that interests me. Of course, the customers have to be interested in what we sell, but we have managed to achieve that. I have shared our financial success with the employees, and we have achieved a strong reputation for innovative and quality work while maintaining positive, ethical values.

ABOUT THE SUBJECT

Richard Lethin is the president of Reservoir Labs, Inc., and an adjunct professor at Yale University. His research focus is the development of software for high-performance computing systems. He has a BS in computer science from Yale and a PhD from MIT, where he was a Hertz Fellow.

7 A BIGGER PICTURE

Jennifer Park

I carefully manipulated the small piece of tubing with tweezers inside the tissue-culture hood. In the middle of the tubing grew stem cells embedded in a collagen gel that had been cultured for a week. Everything had been carefully sterilized—the tubing, the pipe fittings, the tweezers. The next step was to insert the tubing in a small chamber where a regular "heartbeat" from a pulsatile pump would, I hope, coax the cells to turn into the smooth muscle cells of the circulatory system.

A week later I checked the experiment. I could already smell failure. The growth medium was emitting the faint smell of chicken broth, and a bit of mold was starting to grow around the fitting. Somewhere in the painstaking steps I took, I had introduced bacteria. I would have to start all over again, ditch this failed, expensive experiment and troubleshoot what went wrong. Repeat and repeat again. Such was my life in graduate school.

I like variety. I like traveling to new places, trying new foods, and exploring various hobbies. I recently went scuba diving in Cuba and the Galapagos, figured out how to cook a pasta dish I sampled on a visit to Sicily, played clarinet in a woodwind quintet concert, and displayed work at several photography exhibits across the United States. Many different things interest me, which makes it a challenge to focus on one subject in depth for many years. My desire for variety attracted me to Selventa in Cambridge,

Massachusetts, a company that uses computational techniques to under-stand the mechanisms of disease and drug response using a systems-biology approach.

The work at Selventa was very different from the work I did in gradu-ate school—instead of focusing on one disease area, I learned about many. I could see trends across mechanisms in different diseases and appreciate their differences. My work at Selventa felt efficient, wide reaching, and more immediate—we were helping to develop drugs to treat patients in the next few years rather than trying to understand the inner workings of cells for treatments sometime in the future.

I also liked the communication aspect of my job, working with different teams to bring a project to successful completion. In my time at Selventa, I transitioned from individual contributor to research director, oversee-ing collaborations and managing communication between teams. I have recently joined another Boston-area biotech company, where I have moved even further into management as a director of business development. In my new position, I work with clients to understand their problems and goals as well as with the company's internal project teams to ensure we satisfy our clients' needs. I work on a range of diseases, which satisfies my desire for variety, and I continue to learn about new drugs, diseases, and biological mechanisms, a practice that contributes to designing optimal therapies that have a real chance of making it into the clinic.

EARLY LEARNING IN SCIENCE AND HUMANITIES

I grew up in a suburb of Houston called Clear Lake and attended good public schools. Both my parents were employed by NASA contractors at the nearby Johnson Space Center. My father developed guidance navigation-and-control systems for manned space programs, including the Space Shut-tle and the International Space Station. My mother was a software engineer and worked on rocket-engine simulation. Both my parents went to college in Korea and came to the United States for their graduate work. My father studied for his PhD in electrical engineering at Oklahoma State University. My mother was getting her master's in English at Fairleigh Dickinson when she met my dad, but after moving to Houston, she got another master's in computer information systems at the University of Houston.

Learning about both science and the humanities was an important part of growing up in our family. I experimented with a chemistry set, assem-bled an ant farm, grew sea monkeys, and participated in science fairs. My dad would give me and my sister math problems on our way to school and

Jennifer Park grew up in the Houston area, where her parents were employed by NASA contractors. She and her parents visited the Johnson Space Center, circa 1989.

help with science fair projects. I was also exposed to the arts and music through my mom. She gave us classic literature books and poetry anthologies as Christmas presents, alongside the human-body model set from my dad. She took us to classical music concerts and signed us up for music and art lessons. In high school, I was very involved with band and dreamed of playing clarinet in the New York Philharmonic.

When I was finishing high school, my two top college choices were Berkeley and Cornell. I chose Cornell, deciding to strike out on my own in a beautiful rural town rather than move to a city where my sister was already attending university. At Cornell, I took science and engineering courses as well as liberal arts and music. I played clarinet in the wind ensemble and jazz band and wondered what it would be like to have a career as a musician. I felt conflicted about the direction of my career, but as a practical person with a practical upbringing I chose to pursue science, which felt like a more stable and socially impactful choice.

CREATING LIFE

Cornell has an externship program in which you shadow an alumnus at his or her job during winter break. Since I was considering a career in medicine, I observed a plastic surgeon at the MD Anderson Cancer Center in Houston, who would reconstruct patients' appearance after their tumors were removed. I witnessed patient consultations and surgery. Prior to the externship, I hadn't spent much time with sick patients. Watching difficult decisions being made and then observing their outcomes made a lasting impression on me. During the externship, I first heard about the field of tissue engineering. The head of bioengineering at MD Anderson told me about work being done to replace diseased tissue with natural tissue that could be cultured and grown. I found this cutting-edge research very exciting. When I returned to Cornell, I decided that instead of becoming a doctor, I wanted to be a bioengineer and create new organs.

I had three opportunities to do research in bioengineering at Cornell. My first research experience, with Professor Larry Walker, began the summer after my sophomore year. I was accepted into a research program funded by the National Science Foundation and studied how to break down cellulose using cellulase enzymes for waste disposal. The summer after my junior year I found a tissue-engineering research opportunity at Georgia Tech in Dr. Marc Levenston's biomechanics laboratory, where I learned cell culture technique, growing cartilage-like tissue from chondrocytes using different extracellular matrices. And during senior year I was employed as an undergraduate research assistant to Dr. Mark Saltzman on a dental-tissue-engineering project, again working with cells and matrix to create tissue replacements.

As a senior in college, I spent a lot of time playing clarinet in the university jazz ensembles, helping to run the ensembles as well as practicing clarinet, doing research, and studying for classes. At the end of the year, I found out that both a close musician friend and my jazz teacher were moving to the San Francisco Bay Area in the fall, so I decided to attend the University of California–Berkeley for a PhD in bioengineering, where I could study tissue engineering and continue to play music with my friend and jazz professor. In the back of my mind, I was still considering the possibility of becoming a musician instead of a bioengineer.

Berkeley was a great place to study bioengineering; several professors there were working in areas I found interesting. I rotated through Dr. Frank

Jennifer Park worked on tissue engineering as a graduate student at the University of California—Berkeley. She presented stem cell results at a conference in 2002.

Szoka's drug-delivery and gene-therapy lab. If I had been a more independent researcher with strong creative ideas, I would have done well in Dr. Szoka's lab, but since I was not sure what project to focus on, I decided I needed more guidance. I also rotated through Dr. Song Li's lab. Song was working on tissue engineering using stem cells and nanofibers. As a new professor, he was still doing hands-on research in his lab and had time to spend with students. I chose Song as my PhD adviser and became his second graduate student. Song was able to give me the time and attention I needed.

FASCINATING FIELD BUT TEDIOUS EXPERIMENTS

In 2001, tissue engineering was a new field, with the promise of lab-grown biological replacements for diseased organs. Stem cell therapy was also new, with its promise of generating cell types to create organs if given the right environment. Our lab also spun nanofibers consisting of biocompatible materials as scaffolds for cells to align, which is important in blood vessel and neural applications. Combining three futuristic concepts—tissue engineering, stem cells, and nanofibers—my graduate work seemed very cutting edge and exciting. My PhD dissertation focused on differentiating stem cells using mechanical cues to create vascular tissue replacements for atherosclerotic blood vessels.

Although the science sounded fascinating, the actual experiments were tedious, and they often failed. I enjoyed working with my lab colleagues and my adviser but not cleaning up bacterial contamination. The stem cells required continual care and nurturing, and my experimental work seemed to consist mostly of manual labor and troubleshooting. I remember feeling like a plumber, assembling pumps, tubing, and fittings. The best part of my graduate experience was analyzing data and finding trends in the biochemical assays and microscopy images. From this work, I learned that I prefer analysis to direct lab experimentation. After six years, I completed the research and published papers describing how mechanical stimuli induce stem cells to differentiate into different cell types.

Jennifer Park is currently the director of business development at Applied BioMath in Lincoln, Massachusetts. She presented pharmacokinetics-modeling results at an industry conference in 2016.

After graduate school, I decided I wanted to work in industry on practical projects with an end product in mind, but I had a hard time finding an industry job. Most of the advertised positions required a postdoc or industry experience. I eventually found a position at a start-up biotech company in Worcester, Massachusetts. Advanced Cell Technology (ACT) was trying to make blood cells from embryonic stem cells, an area that overlapped with my cardiovascular expertise from graduate school. After working at ACT for a year, I published a paper on differentiating red blood cells from embryonic stem cells. However, the start-up was not doing well financially, and I decided to leave. I was tired of the repetitive experiments and going into the lab every weekend to care for stem cells. It was time to do something different.

A BETTER FIT

Looking for a job on the Monster.com search engine, I selected "PhD" as my level of education and "0–1" for number of years of industry experience. Of all the jobs posted, only one remained that fit these parameters. This position, in a company called Selventa, looked really interesting—an analytical job doing molecular and computational biology, but without any lab work. The projects would allow me to learn about many types of diseases. The company worked in systems biology and focused on the big picture, something that attracted me after years of painstaking focus on experimental details. The job was in Cambridge, the center of the Massachusetts biotech industry. Moving to Cambridge would also make it easier to attend musical events and meet like-minded young people.

However, the job used computer programming to do computational biology, and I didn't have expertise in either one. When I interviewed, though, the staff said those skills could be learned on the job. What they needed was someone with a strong molecular biology background who could read and understand journal articles in biomedical research. I had the necessary background and liked the people and the job, so I joined Selventa as a scientist I.

At Selventa, we created computational drug models for a wide range of inflammatory, oncological, and metabolic diseases such as ulcerative colitis, lung cancer, and diabetes. After the computer made a statistical prediction, we would vet mechanisms to ensure they made biological sense within the context of the disease and built networks using both known mechanisms and novel mechanisms and literature connections.

When I first started at Selventa, I felt lost. I had never done such in-depth molecular biology, and everyone around me seemed so smart and knowledgeable. But I eventually found my way and stopped comparing myself to my colleagues. I began leading projects and teams and having more client interactions. I realized my strength lies in management skills such as internal communication and explaining project results and impacts to clients. Rather than brainstorming novel scientific ideas, my forte is connecting people, seeing the big picture, and using my understanding of the science along with a practical understanding of milestones and timelines to help teams to be effective in accomplishing their goals. I realized it is an art to explain complex results through graphics in presentations and enjoyed combining my artistic and scientific talents.

After seven years at Selventa, I decided to apply my strengths in communication, management, and big-picture understanding to another position in the biotech world. I left Selventa in 2016 to work at another Boston-area biotech-modeling company, Applied BioMath, in a business-development role.

At Applied BioMath, we help pharma/biotech companies better understand their drugs by using mathematical models to identify optimal properties and dosing regimens. Our analyses help speed up the drug-development process by simulating patient outcomes before clinical trials, which enables experiment prioritization. I meet with clients, make sure our project teams' work is aligned with client needs, and seek opportunities for new projects. I still get to do some science and have to understand diseases and drug mechanisms to be effective, but my role is primarily a communicative one. In my new job, I get to see the big picture and learn about all the phases of drug development. I enjoy my job and take satisfaction from helping speed the development of new and better drugs to benefit patients.

ABOUT THE SUBJECT

Jennifer Park is director of business development at Applied BioMath in Lincoln, Massachusetts. She has a BS in bioengineering from Cornell University and a PhD in bioengineering from the University of California–Berkeley.

8 PERSONAL CONNECTIONS AND SERENDIPITY

Stephen D. Fantone

In 2015, I received a surprising and meaningful honor—the University of Rochester's Distinguished Scholar Award. In my acceptance remarks, I acknowledged my early education in the public schools at West Hartford, Connecticut, for setting me on a wonderful career and intellectual trajectory. In particular, I paid tribute to my ninth-grade Latin teacher, Mr. McEvoy. His stern but encouraging instruction provided me with the skills to learn on my own and to invent, create, and later synthesize products, technologies, and businesses as well as to communicate ideas effectively.

Although I took Mr. McEvoy's Latin class nearly 50 years ago, I continue to benefit from his lessons: strive for excellence, seek depth of understanding, respect yourself and others. His class was both a transformative experience and a philosophical framework for future scholarly and business pursuits.

A TELESCOPE IN THE BASEMENT

Growing up in West Hartford, I was a good student and always had a strong drive. I would take five or six classes in a semester, go to summer school, and usually select the most advanced classes. I was interested in astronomy, photography, and electronics. By age 15, I was spending long hours building and using telescopes and their associated mounts, drives, and cameras. I financed these activities by delivering newspapers, cutting lawns, and shoveling snow. Just prior to my junior year of high school, my father took a job with Polaroid

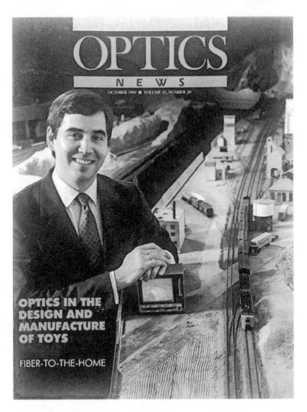

Steve Fantone was featured on the cover of *Optics News* in 1989 for optics in toy manufacture and design.

in Massachusetts. Our family was going to move, so, rather than attend a new high school for senior year, I decided to try to get into college early.

I was fortunate to be admitted to MIT and enrolled in the Electrical Engineering Department. At MIT, I continued the tendency toward overwork that I had begun in high school. In the spring of freshman year, I really overdid it. I enrolled in seven courses: calculus, differential equations, two physics courses, chemical thermodynamics, a humanities course, and Russian. The load was completely crazy, and the stress literally gave me stomach pains.

I visited the medical clinic about a month before the end of the term, thinking that I was getting an ulcer. The doctor asked what was going on in my life. I told him that I was taking seven courses; that my father had had a coronary the previous month and was currently at Massachusetts General Hospital; that my living arrangements were less than ideal; and that final exams were looming. Finally, I told him that I was drinking 15 cups of coffee a day and eating lots of donuts.

The doctor said, "Son, you need to slow down. Eat a real breakfast and lunch, cut out the junk food, cut back on the coffee, and regularize your schedule."

I was successful in passing all my classes, although I certainly would have better mastered each subject had I taken a normal load. I was an immature 17-year-old college freshman and didn't know any better. However, the experience was invaluable as it helped me realize the maximum amount of work I can do at one time. It was an important lesson that helped me know my limits, something I wouldn't have learned if I hadn't pushed so hard.

By age 19, I had earned enough credits to graduate MIT a year early. I was still interested in optics and took courses in electricity and magnetism, but I became frustrated with the overly theoretical orientation of these subjects. I was lucky to land a co-op job at Honeywell Radiation Center that involved the design of space-based optical platforms and provided a practical grounding for my academic studies. I spent a fourth year at MIT taking advanced courses in electrical engineering and completing a degree in management. My interest in business, which had funded my scientific activities when I was a kid, continued during my time at MIT, where I designed and built optical instrumentation as a side activity.

GRADUATE SCHOOL

I had a nice set of options following MIT, with offers from several universities, but my interests were in optics, and so I chose to attend the Institute of Optics at the University of Rochester. During my first year, I was assigned Professor Douglas Sinclair as an interim adviser, which proved fortunate. Sinclair was president of the local section of the Optical Society. We became friendly, and he asked me if I would be his house chairman for the society's bimonthly seminars. He explained that the house chairman arranges for the lecture hall and makes sure there are refreshments afterward. It was an unexciting job, but it turned out to be invaluable for making professional contacts.

Before each dinner, the executive committee and the house chairman would take the speaker out to dinner. As a poor graduate student, I appreciated a free meal at a nice restaurant, but, more importantly, these dinners meant I had the opportunity to meet all the leaders in the Optical Society in Rochester, New York, and the visiting speakers. Pretty soon I had an expansive network of colleagues. They would tell me about interesting research going on in their labs at Xerox or Bausch and Lomb and would offer me tours. That was just one example in my career where volunteering and service opened up exciting opportunities.

During my first year, I was awarded a Hertz Fellowship, which gave me the freedom to choose a thesis area and adviser—independent of funding considerations. I chose Duncan Moore, a young professor working in gradient index optics, as my thesis adviser. I was Moore's first graduate student, and, in fact, he had completed his PhD at Rochester just the semester before I arrived.

One of the keys to success for our company and my career was to become an expert in a technology with broad application across industries. That is different from specializing in a particular industry. We have customers who are in the defense industry but also [in] medical, industrial, and commercial companies. When there's been a downturn in one area, there has typically been demand in another area, and that has allowed for stability and growth.

Because the field of gradient index optics was relatively unexplored at the time, it included many interesting unsolved problems. Gradient index optics uses inhomogeneous lens materials with varying indices of refraction, created by doping glass with elements such as lithium or sodium. Lenses made from gradient optics can be very compact and make possible new devices such as small table-top copiers. I looked into questions of how to correct aberrations and understand the impact of lens parameters and designs on image quality. I calculated the higher-order coefficients of aberrations and created a model that tied the varying glass composition to a corresponding variation in the refractive index and its spectral dependency.

I had a close collegial working relationship with Duncan Moore, whom I met with once or twice a week. Because I had funding from the Hertz Fellowship, I was able to maintain considerable independence. My priority was to learn whatever I could at the Institute of Optics, finish my thesis, and move on with my career. These objectives were not perfectly aligned with those of a particular research group, so the fellowship provided a buffer and some leverage that allowed me to pursue my personal interests and goals.

As I neared the end of graduate school, I considered a number of job options. I was offered a faculty position at the University of Arizona, a research position at a university lab, and two industrial positions, including one from Polaroid. I took the Polaroid offer for the opportunity to work with the highest-caliber people. Within a month of joining Polaroid, I was doing a project with Dr. James Baker, a world-class lens designer, who together with Edwin Land had convinced President Eisenhower to build the U-2 spy plane. I worked on all sorts of fun projects—3D movies, holographic imaging, and fiber optics—and made connections with scientists such as Edwin Land, Bill Plummer, and James Baker. I have continued making connections with wonderful scientist/engineering mentors who became friends throughout my professional career.

A DIFFICULT DECISION

In 1984, as a 30-year old lead engineer at Polaroid I faced the hardest decision of my life. I had received an unsolicited job offer from Milton Chang, president of the Newport Corporation, asking if I would lead its new optics division. The offer would double my salary and include stock options and a company car—a very attractive position. However, for several years I had been weighing the idea of starting my own company and had already incorporated under the name Optikos Corporation. I was trying to determine my career direction, and the stress I experienced was so intense that I became physically ill each morning!

I went to my boss at Polaroid, Bill Plummer, and showed him the offer letter. I explained that I didn't want the corporate job at Newport and wasn't asking Polaroid for more money. What I really wanted was to commit to my own optical consulting company, with Polaroid as my first client. Bill said he would support me, and Bill's boss also agreed. I thought it was a done deal, but then I was told that the head of engineering absolutely refused. Meanwhile, Milton Chang was calling every day, asking whether I would accept his job offer.

I phoned my mother and asked her what I should do. She said, "I don't know the first thing about optics or your career, but whatever you do, it has to feel right in your gut." She was absolutely right.

Through my employment at Polaroid, I had become a friend of Howie Rogers, who was in charge of Polaroid's research-and-development division. During a technical meeting, he sensed something was bothering me and asked about it. I told Howie of the situation and that without a commitment from Polaroid I might not be able to resist the job offer from Newport. Howie went to his phone, called the head of engineering, and said, "If you won't give Steve Fantone a consulting job, then I will; and if I do, you won't get any of his time!" By the time I returned to my office, the deal was set; the contract was approved. A personal friendship got Optikos off the

A serendipitous airport encounter led Steve Fantone (*right*) to a friendship and business partnership with Benthos founder Sam Raymond. Steve was Benthos chairman from 1997 to 2007. Other Benthos board members (*from left*): Ted Mollegan, John Coughlin, Jack Naylor, and Sam Raymond.

ground. And Howie wasn't the only one who helped; there were many others, and this instance was only one of a number of times during my career when friendships and personal relationships have created opportunities and new directions. And these friendships have often started from chance, serendipitous encounters.

A CHANCE ENCOUNTER

My most important serendipitous encounter occurred on March 31, 1986, at 4:00 a.m., when my wife, Betsy, and I were in line at Logan Airport waiting to catch a charter flight to Aruba to go scuba diving. Everyone standing in the line, except for one or two of us, was staring at his or her feet, half-asleep, and wondering why he or she had agreed to take such an early flight. I noticed ahead of me in line a distinguished looking older gentleman who seemed pretty perky for 4:00 a.m. He was wearing an MIT class ring. I am always interested in talking to people, so I asked him the question that MIT alumni always ask: "What course were you?" He answered, "I was course 2," which translates to, "I majored in mechanical engineering."

He told me that he was Sam Raymond, class of 1950, that he had a company on Cape Cod called Benthos, and that it made oceanographic instrumentation. He told me about its products, including deep-sea releases, clear spherical enclosures to protect and provide constant buoyancy for cameras and equipment placed in the ocean, and underwater-camera housings.

I introduced myself as a member of the class of 1974 and mentioned that I also had a company with a Greek name involved in optical engineering. In Aruba, Sam and I went scuba diving together and became fast friends.

Six years later Sam Raymond asked me if I would like to join a group called the Hamilton Trust. The Hamilton Trust is the oldest private investment club in the United States, currently with about 80 members. Sam Raymond had been sponsored into the club by Harold "Doc" Edgerton,

Steve Fantone (*left*) and his PhD thesis adviser, Professor Duncan Moore, attended the graduation ceremony at the University of Rochester in 2017, when Steve received the University Distinguished Scholar Award.

beloved MIT optical engineering professor, who had invented the strobe and was one of the original investors in Benthos. Sam was one of very few technically trained members of the club. By sponsoring me, Sam was continuing an optical engineering legacy. I became a member of the club.

A few years after that, Sam asked me to join the board of directors of his company, and when he decided to step down as chairman, he asked me to be the new chairman. I was chairman of Benthos for ten years before we sold Benthos to Teledyne Technologies in 2007. The chance meeting at Logan Airport had led to a completely unexpected set of career opportunities for me in the field of optical engineering and instrumentation—a field that had captivated me from an early age—and to a lifelong friendship with Sam Raymond.

VOLUNTEERING

Since my graduate school days, I have continued my involvement with the Optical Society. I have held volunteer roles of increasing responsibility from technical program chair to treasurer and vice president. Serving in those roles put me in close contact with the leading optical researchers and kept me updated on all the latest developments in the field of optics around the world.

From my work at the Optical Society, I also learned about the governance of not-for-profit organizations and how to help those organizations succeed. I have been able to apply that knowledge in my role as a board member of the Hertz Foundation, as the chairman of the Pioneer Institute, and in my service on other boards, including both for-profit public companies and nonprofits. I have been involved in activities I never contemplated in graduate school, ranging from inventing and licensing toy ideas to public-company proxy fights as well as corporate mergers and acquisitions. The Hertz Foundation encouraged me to think big, and I remain grateful for the support and encouragement that the Fellowship Award and the exceptional individuals in this organization provided.

In parallel with all of these activities, support from my family and in particular from my wife, Betsy, was instrumental in my taking on all these challenges. I was extremely fortunate in finding my wife—her help, encouragement, and love have provided the support and confidence that have allowed me to be engaged in a broad range of activities and provided a true balance to my work life. I cannot overstate how much she has enriched my life.

ABOUT THE SUBJECT

Stephen D. Fantone is the founder and president of Optikos Corporation and a senior lecturer in the Mechanical Engineering Department at MIT. He has undergraduate degrees in electrical engineering and management from MIT and a PhD in optics from the Institute of Optics at the University of Rochester.

9 TANGIBLE PRODUCTS

Kathy Gisser

I found my first adviser and first chemistry job thanks to an auspicious alphabetical opportunity.

Laurel School, the all-girls school I attended, allowed seniors to do full-time projects for the last six weeks of the year. At the end of my senior year, I knew I would be faced with the prospect of watching movies in class after taking my advanced-placement exams and getting accepted to college, so I decided to find something more productive as a senior project. At the end of my summer in the biology lab, I picked up a brochure from the Chemistry Department at Case Western University and started leafing through the professors' research descriptions, hoping to find one whose research matched my interests.

I quickly gave up trying to understand the jargon-filled descriptions and instead decided to contact each professor, starting at the beginning of the alphabet and moving through the list. That's how I found myself in Dr. Al Anderson's office—he was the first professor listed in the directory. He asked me about my math background, and I explained that I had taken trigonometry and would be taking calculus and that I could provide six weeks of free labor, continuing longer if he would offer me a paid summer job. And it worked! Al hired me—not just for that summer, but for several summers.

That entrepreneurial initiative has helped me find jobs from that first internship in high school through my current positon as a staff scientist at Sherwin-Williams. It served me well when I was in school and later in my work life, allowing me to be part of fascinating projects with mentors and colleagues I wouldn't otherwise have met. The more I asked, the more opportunities I had and the more confident I became. This confidence would help me overcome the obstacles I encountered along the way.

As I learned and grew, I found that I preferred working in large cross-functional teams, where my talents could be combined with the skills of my teammates to tackle complicated problems. The challenges I enjoyed most were those focused on making a product, something tangible that people would want and find useful. During my career, I have participated in the development of several significant products. Once I figured out that I prefer to work with and lead teams, amazing things began to happen.

AN ENTREPRENEURIAL SPIRIT

I was born outside of Cleveland, in Shaker Heights, Ohio. My dad was a pharmacist. Watching my parents run their own business showed me, before I was even old enough to realize it, what an entrepreneurial spirit was. I learned the basics of the business from helping out—tasks such as keeping track of inventory and serving customers—but I also learned about having new business ideas and following through until they came into being.

My mother taught middle school before I was born and then taught herself how to do the accounting for my dad's store. She nurtured my love of science from an early age. In elementary school, I jumped at every opportunity to do a science-related activity. Long before I knew about product development or materials research, I liked the "stuff" of science. I liked working with tangible things and was fascinated by the mysteries of materials that did one thing when they were alone and a different thing when they were mixed together.

After the first two weeks of chemistry in high school, I knew I wanted to be a chemist. My teacher, Mrs. Daley, was among the best chemistry instructors I have had. Her classes were clear and well organized. When both of us realized that I had a strong interest in the subject, Mrs. Daley asked me to help her with lab preparation for her other classes. I was hooked.

My first opportunity to work in science outside of school came in the tenth grade when I volunteered at the local hospital, working on a project run by one of my classmate's parents. I spent most of my time mixing solutions and running a spectrometer. It was basic work, but I loved the environment and knew I wanted to spend more time in a lab.

The next summer, before my senior year, I applied to a program at Case Western that placed high school students in biology labs. I sacrificed a rabbit the first week there and spent the rest of the summer preparing muscle protein samples with minimal guidance from the other researchers. I enjoyed my independence and the thought that my contribution was going to support the work of the graduate students, but I didn't enjoy biological lab work as much as doing chemistry at the hospital. As a result, I decided to do chemistry instead of biology for my senior project.

Because Dr. Anderson was the first professor listed in the Chemistry Department directory for Case Western, I found my way to his office and introduced myself. He asked how my trigonometry was, and when I replied that it was fine and that I would be done with calculus by the next spring, he told me I was hired.

It turned out that trigonometry was the right skill set for the molecular orbital calculations that Dr. Anderson and his graduate students were performing. He provided me with chemical structures, and I arranged them in three-dimensional space, using trigonometry to find their positions. We sent my calculated positions to a Cray computer at NASA, which computed the shapes and sizes of the molecular orbitals. I didn't yet know enough chemistry to use the results, so I would make a record of any low-energy configurations and give that record to Dr. Anderson to analyze.

I grew to love the lab atmosphere, finding my own niche among the graduate students and other researchers. I can't overstate how much it mattered that I was respected and treated as a contributing member of a research lab even before I started college. Based on that experience, I knew I was suited to do research and to continue in science through graduate school if I wanted.

CLASSICS AND CHEMISTRY

For college, I was interested in a school that had great courses in both the humanities and the sciences, and my college councilor suggested Yale. The cheerful atmosphere on campus when I visited in the spring confirmed to me that I would like it. I enjoyed both my social life and the courses at Yale and have lifelong friends from my time there. I already had an interest in the classics, and I continued to study literature from the Greeks to Shakespeare. In the end, I graduated with a double major in classical civilization and chemistry. I took the required math classes but was more interested in math as a tool than as a subject in its own right. I did enjoy a Chemistry Department course in group theory, learning the symmetry groupings that are important to understanding the electronic structure of atoms and molecules. It reminded me of the work I had done in Dr. Anderson's lab.

The chemistry faculty was enthusiastic about my prospects as a chemist. Professors in both science-technology-engineering-math and humanities courses gave me good advice and encouraged me not to give up because of temporary challenges. In my sophomore year, I decided to take "physics for physics majors" because I had heard great things about Professor Zeller. However, I got a D on the first semester midterm, and the course grade was based only on the homework, midterm, and final grades. I went to Professor Zeller's office hoping for a way out that didn't involve dropping the class. I will remember his advice forever—he said that every year a few students are completely wiped out by the midterm, but I should keep at it because most of them figure it out and do well on the final. He asked me if I had a good group to do the homework with because the homework was "too hard to do as an individual" and advised that I should take the risk that I would be able to do well on the final. He was right; I was able to "figure it out" and do well. And working on homework for his class with a group introduced me to the idea that a group could tackle and solve problems that would be too hard for one person to solve.

Entering Yale, I briefly thought about pursuing chemical engineering but heard that the department wasn't friendly to women, so I decided to stick with chemistry. But in my senior year I did a very enjoyable senior research project on colloid stability mentored by a chemical engineering professor, so perhaps I made that decision without gathering enough information.

There was only one truly discouraging experience while I was in college. During sophomore year, I found myself struggling in the advanced organic chemistry course. At the end of the term, I had a C+ in the class. Along with the poor grade, the professor also gave me the unsolicited advice that I shouldn't pursue chemistry as a career. One of my male friends who got the same grade wasn't offered the same advice. I was disgusted with the professor's attitude. I knew from my work with Dr. Anderson that I could succeed in chemistry research— because I already had!

Through my college years, I continued to do summer work as a researcher. After my freshman and sophomore years, I worked for Dr. Anderson. After my junior year, I participated in the Bell Labs Summer Research Program for Minorities and Women, which led to my first publication. In the summer before I started graduate school, I was hired by Yale's Department of Chemistry to teach in the Yale Summer School. When I spent sleepless nights wondering if I had what was needed to become a good scientist, I could look at my track record and see that I was already a success.

Kathy Gisser and her future husband, Dan, studied chemistry as undergraduates. They attended the Yale prom in 1987.

GRADUATE SCHOOL FOR TWO

Throughout college, I had continued dating my high school sweetheart,
Dan, who was attending Dartmouth College. By the end of junior year,
with both of us looking at grad school in chemistry, we decided that we
should find somewhere we could go together.

So the summer before senior year we spent a week in a friend's cabin
in rural Wisconsin with a stack of graduate school brochures. We looked
for departments that had enough physical chemistry for Dan and enough
materials chemistry for me. We also wanted a department that was large

enough so that if we broke up as a couple, we could still be comfortable staying there to finish our theses.

Five schools fit the bill. The University of Wisconsin accepted both of us, so we went there.

After a year of course work at Wisconsin, I chose Professor Art Ellis as my thesis adviser. Half of his group did photoelectric chemistry, and the other half worked with uranium salts and other glowing materials. One day Art came to me with a length of metal wire in his hand and told me to listen carefully. He dropped the wire, and it hit the ground with a thud. He picked it up, ran some hot water over it, and dropped it again. This time it rang like a bell. That's all it took to know I wanted to understand that material and work with it for my thesis.

The wire was a nickel-titanium alloy and underwent a crystallographic transition when it was warmed to just above room temperature. In the high-temperature phase, it propagated sound waves surprisingly well. But Art wasn't interested in sound. He said, "I think if we could get thin films of this onto a semiconductor, we could do some interesting photochemistry as it goes through the phase transition."

Although I chose Professor Ellis as an adviser because of our shared technical interests, it turned out that our working styles were not really compatible. Art was a hands-off adviser, which many of his students appreciated, but I wanted mentoring and a closer relationship with my adviser. Fortunately, Art encouraged collaborations with industry, which was unusual in the early 1990s. My industrial collaborators filled in some of what was missing in my academic environment. Nickel-titanium materials had commercial applications, making the project a good choice for collaboration with industry.

I read up all I could about nickel-titanium, and at the end of an article in *Popular Science* I found a list of three companies that worked with the alloy. I called them all, asking about thin films, and connected on the third one. TiNi Alloys, a small firm in Oakland, California, had just purchased a vapor-deposition system to make thin nickel-titanium alloy films.

TiNi was a very small company with only four employees: the founder, an engineer, a machinist, and the office/marketing manager who handled the company's finances. I worked very closely with the founder, Dr. A. David Johnson. He became as much of a mentor as Art, and working with TiNi was my first exposure to the joy of working to make a tangible product.

My work with TiNi alloys also caused me to change the goals of my thesis. Although I had originally hoped to order samples for my own photo-electrochemical experiments, I became totally invested in working with the team at TiNi to figure out the deposition process. We applied and received a National Institutes of Health Small Business Innovation Research grant to use the films as an implantable insulin pump, but in the end we refocused to make microvalves. Our first films, deposited onto room-temperature substrates, were amorphous. However, films deposited onto heated substrates at the proper temperature and cooling rates exhibited the shape-memory phase transition. When we finally succeeded in producing a sample with the desired material properties, I was ecstatic. In fact, I still have that sample as a souvenir.

When I started graduate school, I expected that I would have an academic career. I would complete graduate school, find a postdoc, and eventually become a professor. However, during the last months of my work with TiNi, my perspective changed.

After visiting what should have been my dream postdoc lab at the University of California–Santa Barbara, I realized how much fun I was having in the product-focused, collaborative environment at TiNi up the coast. I realized

Kathy Gisser and her husband, Dan, worked at Kodak in Rochester, New York, in 1999.

that I preferred working in cross-functional teams, where my talents could be combined with the skills of my teammates to tackle complicated problems. I was reminded of my experience in college physics, where in working as a group, my schoolmates and I could solve problems that were too hard for any individual to solve. And the challenges I enjoyed most were those focused on making a product, something tangible that people would want and find useful. So I stopped applying for postdocs and started applying to industrial labs, where I could work as part of a collaborative team.

PRODUCT DESIGN

I spent the winter of 1992 visiting industrial sites around the country. My first job offer was from Kodak in Rochester, New York. By the time I got the offer, Dan and I realized that the job market for industrial chemists was weakening. We decided that I should accept what looked like a fantastic job for me, even though it was not certain that Dan could find work in the Rochester area. I took a long break before starting work at Kodak, during which we got married. Dan joined me in Rochester to finish his thesis. We decided it was most important that we live in the same city and trusted that Dan would find a job within commuting distance of Rochester after he got his PhD.

When I joined Kodak, I received a degree of training that is almost unheard of today—18 months of training, including rotations in different parts of the company. I essentially worked as a postdoc in areas of photographic science specific to Kodak and not taught in any academic classroom. During my training projects, I was assigned two mentors with different management styles. The relationships I formed with more senior researchers were ones I would rely on for technical and personal advice throughout my career.

By the 1990s, the photographic industry was mature, with well-established products and markets. Although there were still positions in materials research at that time, once I finished my training, I asked to focus on product development, to formulate films and processes that would meet customer needs. I worked with marketers, manufacturing engineers, and other scientists to understand how to balance cost, manufacturability, and our customers' needs. I helped optimize a new movie-film process for a new print film to let cinematographers achieve darker blacks than we believed possible. That film went on to win a Scientific and Engineering Award from the Academy of Motion Picture Arts and Sciences. I wrote new movie-film-processing instructions to reduce water usage. When a colleague came to me with a problem in our safelight coatings, I led a project to reformulate them and our color-filter products for the first time in a century. Each of these efforts

Kathy Gisser and her husband, Dan, moved back to their hometown, Cleveland, Ohio, in 2006. Since 2007, she has worked as a staff chemist at Sherwin-Williams. Most recently she was the program leader for PaintShield, the first microbicidal (bacteria-killing) paint approved by the US Environmental Protection Agency.

was a large collaborative development project in which up to 50 specialists from across the company contributed their expertise and jointly agreed on next steps. Each lasted between two and five years, with a budget ranging up to several million dollars a year.

One consequence of relocating based only on my job offer was that Dan struggled to find the right job. A year after we moved to Rochester, he began working at 3M in Buffalo, more than an hour's drive from our house and even longer in the winter. That commute eventually became too wearing, so Dan decided to go back to school and earned an MBA at the University of Rochester. He then found a job closer to home, also at Kodak. A few years later we brought our daughter home, which made it more important that our whole family continue to live together in the same city.

RETURNING HOME TO CLEVELAND

When our daughter was four, Kodak's downsizing began to accelerate. Dan and I always felt relatively secure when layoffs were based on performance, but now the company was cutting deeply into its workforce based on the products an employee worked on. We didn't want an unexpected layoff to force us to move, so we decided to take control of the timing.

Since my career had determined that Dan would move to Rochester, we agreed that our next move would be based on his. We hadn't planned to move back home, but the first great job Dan found was at Eaton Corporation, headquartered in downtown Cleveland.

I was in the middle of the safelight and filter project, so I stayed in Rochester for two months after Dan moved to Cleveland to start work. I was present for the final project review, and it was very satisfying to see the project ready to launch. Then my daughter and I moved to Cleveland. I spent the first months there getting her established in school, networking, and looking for a job.

One of the informational interviews I had was with Dr. Kent Young, another veteran of the photographic industry, who was now a director at the paint company Sherwin-Williams. Shortly afterward, I saw an online ad for a job at Sherwin, looking for a chemist with microbiology training. Although I didn't have the right technical background, I had good scientific and management skills. I called Sherwin's research-and-development human-resources department to see if I should apply and found out that my résumé had already been submitted for the position.

For my interview, I was asked to give a presentation, so I studied up on paint technology and found that the product-formulation skills I had acquired at Kodak were applicable to all kinds of formulation products: rheology, concentrated polymer solutions, and dispersion of particles and additives into those solutions. In addition, I had skills in managing "phase-gate" projects, which is the rubric that most companies use to lower risk and facilitate the development and commercialization of new products.

Sherwin-Williams wanted me to lead the development of a product, now called "Paint Shield," that would become the first microbicidal (bacteria-killing) paint approved by the US Environmental Protection Agency. The project required all kinds of technical expertise that Sherwin didn't have, so I started to identify external partners for those capabilities. I oversaw the external microbiological research and eventually managed 12 external partners, who did everything from testing to developing new materials. I also coordinated research with our marketing department—for example, finding out how quickly customers would need the paint to kill bacteria and how long they would want that protection to last. When the project moved to the commercialization stage, I supported our product-commercialization team and the project microbiologist.

After several years of hard work and validation testing, Paint Shield is now a product sold through Sherwin-Williams retail stores. Hospitals and dentist offices now have microbicidal paint that can be used on their walls, and its use has expanded to include other places, such as gyms and even prefab military latrines! Since that project, I have been working on other technology-development projects at Sherwin-Williams using the skills I have developed over my career to bring innovations to the market. At Sherwin-Williams, I found a place for my skills and talents, led a team, and created something useful. I couldn't be more proud of the work we have done.

ABOUT THE SUBJECT

Kathy Gisser is a staff scientist at the Sherwin-Williams Company in Cleveland, Ohio. She earned a BS degree from Yale and a PhD from the University of Wisconsin at Madison, both in chemistry. Prior to working at Sherwin-Williams, she worked as an industrial chemist at Eastman Kodak in Rochester, New York.

10 FROM HEALTH CARE TO ROBOTICS

Daniel Theobald

"You already have a successful company, a wonderful wife and kids," my adviser said. "Why exactly do you need to get a PhD?" Between raising two kids at home, working at a growing start-up, and studying to pass the MIT mechanical engineering qualifying exam after a six-year break from school, I was completely fatigued. All of it was more than any one person could do.

My wife, Deborah, and I had moved to Maryland in 1999 when the MIT Space Systems Laboratory was relocated to the University of Maryland. While she finished her master's degree, we founded VECNA on $5,000 from our bank account. A couple of pieces of advice for graduate students—if you have the intention of getting a PhD, it's good to get your qualifying exams over before you take time off. But, more important, you should do what you are passionate about. Some believe in suffering for years in a PhD program to reach a career goal. I believe that happiness and fulfillment should be one of your major career goals. Another major goal should be to see how much good you can do in the world.

I grew up in San Jose, California. I rode my 10-speed bike all over the valley, which was mostly open fields at the time. But it was already a center of technology development, well before it became Silicon Valley. I enjoyed building cars and motorcycles, taking apart robots, and assembling computers. I built an Apple II computer from parts I found in a dumpster. Growing

Daniel Theobald and his team at VECNA designed a prototype battlefield-extraction-assist robot to locate, lift, and carry wounded soldiers out of harm's way.

up there gave me some unique opportunities—for example, I represented the state of California at a supercomputing program at Lawrence Livermore National Laboratory in 1988.

My family was not really focused on college, and I considered not even applying, but I had a friend a year younger than I who was looking into colleges. He was interested in going to MIT and challenged me: "You would never get in there." He showed me a brochure, and the school seemed really cool, so I applied and was accepted but was a bit disheartened when I arrived. I expected to find a majority of students would be as passionate about engineering as I was. I knew it was my calling in life, but many students were still trying to find a direction.

I started in the Electrical Engineering and Computer Science Department. At the time, the MIT computer scientists were concentrating on compilers and microprocessors, but I wasn't interested in building a better computer; I just wanted computers to do things with. So I joined the Mechanical Engineering Department, took all the programming and electronics courses, and got a good education in robotics.

My decision about where to go to graduate school was not made very thoughtfully. My adviser told me I could stay at MIT and write one thesis for both my undergraduate and master's degrees. So I stayed and worked on the Mars Rover program. I met Deborah at MIT, and we got married in 1986.

FINDING MENTORS

One of my failings over the years has been not to seek out really good mentors. But during my student years, I had two mentors whom I remember. Professor Rohan Abeyaratne, an expert in mechanics and materials, was my undergraduate adviser at MIT. I also learned a great deal from Professor Woodie Flowers, with whom I worked on my mechanical engineering Capstone (thesis) project. Professor Flowers taught engineering design and

Daniel and Deborah Theobald were honored at a VECNA company celebration in 2013.

product development and was an early robotics pioneer. I remember in the design-projects class, he told us that we were not allowed to do any welding—an MIT fire-safety rule because it was just too dangerous. Then he said, "Therefore we must close the door," and taught us how to weld. Woodie had an intuitive sense for materials, energy, and batteries and passed that along to me. He would ask us fundamental questions such as "Do you understand how an internal-combustion engine really works?"

There was not a great deal of mathematical rigor in my independent-study program. I suffered through the math in my undergraduate and graduate courses. It was not by plugging numbers into equations but through hands-on trying, touching, and building that I gained an intuitive understanding of systems. I think through problems by writing out a description and solution in prose, longhand—this is how the system will develop. My professors liked that approach, and it has helped me—it is how I lead my teams on engineering projects today.

CREATING COMPANIES

I have created many companies—VECNA actually consists of five companies combined into a single entity. Unlike many serial entrepreneurs, I have not been interested in or motivated to create a company simply to sell it and exit. VECNA has some of the world's most talented engineers, and we believe we can have a lasting, positive impact on the world.

Our company is heavily focused on research and development, providing products and services in areas from health-care software to robotics. For example, we are a leader in patient self-services. For example, we are a leader in patient self-services, providing hospitals with online booking and self-service kiosks similar to those used in the airline industry to book flights and check in at the airport. We also have developed mobile robots that transport materials in hospitals, and larger robots that can transport a wounded soldier to safety.

One unusual aspect of VECNA is the relatively high percentage of our resources we try to give back to the community. Every staff member is paid 10 percent of his or her time for pro bono community-service work. Most of my personal money goes to VECNA Cares, the nonprofit arm of VECNA Medical, which donates technology to the developing world for free. For example, when the Ebola crisis was at its peak, we sent robotic technology to Sierra Leone to assist the medical staff there.

When I started VECNA in 1996, it was harder to start a company—the information on how to do so was not as readily available. Today it is

A collection of VECNA robots gathered behind the company's headquarters in Cambridge, Massachusetts.

easier than ever to start a company. The process is well documented on the Internet—you can start a company in 20 minutes! Also, the risk is low for talented people in technology. If the enterprise doesn't succeed, there is little risk of not being able to get a good job afterward.

There are many start-ups now in the robotics area, but I find most of them discouraging. Most people from an academic environment have not been exposed to the industrial problems that need to be solved, so they just invent yet another telepresence robot or a useless cute robot. Yet there are so many unmet market needs. For example, the CEO of Gillette told me that the company needs to increase the output of its production line by 60 percent, a great application for robotics and technology. This is a problem that needs creative humans to solve; it is not going to be solved by existing technology.

Another typical problem with robotic start-ups is that they try to reinvent everything, so getting the hardware working takes far too long. Start-ups can benefit by using common platforms and technology. If they could band together, they might increase their purchasing power. But, most important, they need to be connected to the people in industry with real needs.

For these reasons, some of my colleagues and I helped established the Mass Robotics Initiative Hub. Mass Robotics is a nonprofit collaborative space in Boston that houses not only early-stage robotics companies but also more established companies. It provides the forum for robotics companies to share resources, for early-stage companies to connect with potential customers, and for people who want jobs in advanced manufacturing to get training.

ABOUT THE SUBJECT

Daniel A. Theobald is the founder and currently the Chief Innovation Officer of VECNA Technologies, Inc., of Cambridge, Massachusetts. He is a serial entrepreneur and inventor in the areas of robotics, software architecture, and artificial intelligence, with BS and MS degrees in mechanical engineering from MIT.

11 ADAPTING AND CREATING

Daniel Goodman

On February 21, 1986, the $370 million Magnetic Fusion Test Facility (MFTF-B) was dedicated at the Lawrence Livermore National Laboratory (LLNL). On February 22, it was shut down before a single plasma shot could be fired. The US Department of Energy had decided to concentrate its resources on toroidal (tokamak) systems, which were showing more promise than linear magnetic mirrors such as MFTF-B.

At the time, I was a fourth-year graduate student at MIT, part of a large research group operating two magnetic mirrors. Shortly after the LLNL facility was shut down, our main experiment also lost funding. I continued doing basic plasma physics experiments and completed my thesis a couple of years later. But the experience of an abrupt change in government research direction made me question my career assumption.

I had expected after graduation to be employed as a government-sponsored researcher at either a university or a national laboratory. But after watching my adviser and scientists from our research group leave their jobs at MIT to create companies, I decided that working in the commercial sector was more attractive and less risky than being dependent on government funding.

Yet the end of our research program at MIT and its talented scientists being forced to become creative entrepreneurs turned out to be very good for the Boston economy. Some of the Boston area's most successful high-tech companies, including the one I currently work for, were founded as a result of the events in 1986.

Instead of a career as a scientist developing fusion as an energy source, I have applied my plasma physics training to solving problems in the commercial sector. I have worked in small and large companies, doing research and developing products for the aerospace and semiconductor industries. And I have found commercial product development to be as scientifically challenging and rewarding as I suspect my original career in government-sponsored energy research would have been.

PHYSICISTS AS HEROES

I grew up in Utica, a small city in upstate New York, and attended excellent public schools. My father was an electrical engineer, and my mother a stay-at-home parent. Growing up, I was a well-rounded student with an aptitude for math and science, but what I enjoyed most was playing musical instruments. My first instrument was piano, followed by the clarinet and the cello. My musical interests were strongly encouraged by my parents, who wanted my brother and me to have musical training and opportunities they had not had when they were growing up.

In high school, my two strongest interests were science and music. As a junior in high school, I took physics and fell in love with the subject. I liked being able to predict how the world might behave based on a few basic principles and equations. During senior year, I continued to study physics on my own and decided I wanted to become a physicist like the scientists who helped us win World War II—Oppenheimer, Fermi, Einstein, and their colleagues. I was also spending a great deal of time practicing the piano and cello, playing in local orchestras, and performing often on both instruments.

My parents urged me to study engineering so I could always get a job. Their advice seemed sound, so I looked for a school with a good engineering department that was also strong in physics. I was accepted at Princeton and majored in electrical engineering, but I also took many physics classes and completed the requirements for the engineering physics program. At Princeton, I also maintained my complementary interest in musical study and performance, which has continued throughout my life. I took lessons on piano and cello, played in the university's orchestra, helped found a string quartet, and performed solo recitals on piano.

The oil crisis of 1979 caused by the Iranian Revolution occurred during my sophomore year. The price of crude oil doubled, and long lines appeared at gas stations. The oil crisis sparked much government investment in energy research, including a significant increase in funding for fusion energy.

The largest fusion research center in the country is the Princeton Plasma Physics Lab. Fusion was characterized as a clean and inexhaustible energy source and often discussed on campus. I decided that I wanted to pursue fusion energy for my career. During senior year, I took the graduate course in plasma physics and did research at the lab, calculating the stability of a twisted magnetic-confinement device called a stellerator. I also applied to the Hertz Foundation for a graduate fellowship. My interviewer was Lowell Wood, a polymath protégé of Edward Teller. Lowell baited me, saying that fusion energy would never become a practical energy source, and we got into a heated argument. I worried afterward that I had forfeited my chance for a fellowship by being disrespectful, but the argument may have instead convinced Lowell that I had the commitment needed to complete graduate school and become a productive scientist.

For graduate school, my two best choices seemed to be either to stay at Princeton or go to MIT. I decided that I wanted to live in Boston, where the social scene would be much better than in the small town of Princeton. In

Dan Goodman defended his PhD thesis at MIT in 1989. He later worked for several companies founded by his thesis adviser, Richard Post.

the spring of my senior year, I visited MIT to find out about research projects and to find a thesis adviser. Most of the staff at the MIT Plasma Fusion Center were working on toroidal tokamak experiments and theory, but there was also a new group headed by a young researcher named Richard Post. He was putting together a team of 100 scientists and engineers to build a large magnetic mirror reactor named TARA, after a Tibetan Buddhist deity whose followers develop their spiritual soul traits, seeking to understand secret teachings. I remember my student colleagues joking that we preferred linear pastries to toroidal jelly donuts. This seemed like the right group for me to join.

A LIFELONG MENTOR

I asked Dick Post to be my adviser, and he turned out to be quite a good choice. Dick was full of creative ideas. As his student, I realized that a key part of my job was to figure out which ideas were winners and to do so as quickly as possible. After I graduated, Dick continued to follow my career, provided advice when I founded a company, and invited me to join him at two of his start-up companies.

During my first two years of graduate school, I held summer internships at the Oak Ridge National Laboratory in Tennessee and LLNL. These jobs gave me a feel for my planned career as a government-sponsored research scientist. They also gave me the opportunity to see what living in the South or in California might be like. I remember thinking how kind, welcoming, and respectful the staff at both laboratories were toward a young graduate student. They also appreciated that I could contribute to the local musical scene. I got to play in the cello section of the Oak Ridge Symphony Orchestra for a performance at the World's Fair in Knoxville and played chamber music with excellent musicians in Livermore. After my second year, I didn't seek out additional jobs or internships in order to focus on my thesis work.

For my thesis, I did plasma physics experiments in a vacuum chamber surrounded by a magnetic coil shaped like the seams of a baseball. This shape created a more stable plasma with better confinement than a magnetic mirror made from solenoid coils. For my first experiment, I tried to manipulate the plasma using low-frequency electric fields and then looked for resonances that I predicted would occur between the linear and rotational plasma motion. I never saw that effect but learned a great deal about plasma diagnostics and how to build radio-frequency generators and ion sources. Nature also didn't cooperate with my second idea for a thesis. I eventually realized the data I had collected could be used to determine some basic plasma properties, including the radial ion transport rate, and so that topic became my thesis. Perhaps the most important concept I learned during graduate school was how long to work on an uncooperative experiment before deciding that I should try something else.

During my third year of graduate school, I received a call from a recruiter working for Goldman Sachs. The company was looking for scientists and engineers to apply quantitative methods to financial modeling and was offering large starting salaries, with the possibility of a lucrative career in finance. I told the recruiter that I was dedicated to finishing my PhD thesis and didn't want to change fields. My brother and college roommate received similar calls and agreed to join Goldman. They went on to successful careers on Wall Street, whereas I continued on my path to becoming a scientist.

After I completed my required courses and passed the qualifying exam, I spent the rest of my time working in the lab. In retrospect, I wish I had taken some courses in microelectronics. I didn't realize it at the time, but that would be the area where my colleagues and I would spend most of our careers. The Chemical Engineering Department at MIT offered a plasma chemistry and processing course, and the Electrical Engineering Department offered a hands-on microelectronic fabrication course. The semiconductor

Dan Goodman performed a piano recital at MIT's Killian Hall in 2006.

industry, which was just taking off, would have far more jobs for plasma physicists than fusion-energy research ever would.

As I neared graduation, I looked for local jobs where my experience in building high-power radio-frequency systems would be an asset. I joined a small contract research-and-development firm named Science Research Laboratory (SRL) and worked on solid-state pulsed-power drivers for lasers and electron beams. There I met my second mentor, Dan Birx. A wiry, athletic guy with a mop of unkempt hair and the ability to do just about anything, Birx could build a novel machine for harvesting walnuts or a new style of personal airplane. He was also a legend in the pulsed-power business, having created the largest induction linear accelerator in the world at LLNL.

SRL licensed the Birx designs to a start-up named Cymer, which made excimer lasers for semiconductor processing. Cymer replaced thyratron tube switches with solid-state devices, thus greatly increasing reliability. It

Dan Goodman and Hertz Fellow Jennifer Schloss attended a symposium celebrating the fiftieth anniversary of the Hertz Foundation in 2013.

submitted an initial public offering and became the dominant company in the field.

I learned a tremendous amount from Dan Birx—practical things such as all the uses for a particular style of ignition pliers and how to envision, fabricate, and assemble very complex systems. One area in which I have tried to surpass Dan is reliability testing. Unfortunately, the small plane that he designed was not sufficiently reliable. It crashed during an emergency landing in March 1999, killing Dan and his wife and robbing the world of an amazing and productive genius.

The work at SRL was hands on and in small teams. This was a good fit for me. I have always liked being in the lab, building and testing equipment. I like projects of moderate complexity and duration, and I have little patience for bureaucracy. My graduate work at MIT was good preparation for SRL—in both cases I had to figure out how, with minimal oversight and limited resources, to build systems that were good enough.

STARTING A COMPANY

One of my jobs at SRL was to find new uses for the company's high-energy electron-beam accelerators. A promising area was the curing of composite resins and adhesives, which could be done quickly and at room temperature. This was useful for manufacturing in the aerospace industry. I licensed the technology from SRL and started Electron Solutions Inc. to sell the equipment and the specialty materials. We sold to NASA and Boeing and the US Air Force. In the process, I learned many things that aren't taught in graduate school—about sales and contracts, delivery and service, payroll and accounting, and personnel management. It was a great introduction to the capital-equipment business, which is a tough business—especially if you are a small company.

After my wife and I had our first daughter in late 1999, I decided to change the work–family balance by devoting more time to family and less to business. Dick Post, who had been following my progress, offered me a choice of product management or research jobs at Applied Science and Technology (ASTeX), a company selling plasma-based processing equipment to the semiconductor industry. I preferred to work as a research scientist developing new products.

In 2001, a year after I joined ASTeX, the company merged with MKS Instruments of Andover, Massachusetts. Most of the ASTeX products were smaller components that were sold to large semiconductor capital-equipment manufacturers to be incorporated into chip-manufacturing equipment. The one exception was a physical vapor deposition (PVD) sputter tool used in the final packaging steps of chip manufacture. The ASTeX PVD system competed with one made by Applied Materials. Applied is MKS's largest customer, but it is not a good business plan to have a product line in competition with your customer, so the ASTeX PVD group was spun out during the merger and became NEXX Systems Inc., with Dick Post as president. In 2004, I joined Dick at NEXX, this time as an engineering manager. I learned about documentation and formal engineering processes, another discipline that is not taught in graduate school.

In 2008, Dick Post retired from NEXX, and Tom Walsh took over as president. Tom brought with him a host of experience from his days as an executive at IBM and Novellus. He took the company to a new level of performance, leading to our acquisitions by Tokyo Electron in 2012 and by ASM in 2018. My primary tasks now are directing new product development and overseeing the company's intellectual property. I still try to spend some time in the lab because hands-on research is something I have enjoyed since my graduate student days. And I still enjoy studying and performing classical and folk music, which together with science has been my lifelong passion.

ABOUT THE AUTHOR

Daniel Goodman is director of advanced technology at ASM-NEXX of Billerica, Massachusetts. He has an electrical engineering degree from Princeton University and a PhD in plasma physics from MIT, where he was a Hertz Fellow.

12 MY DREAM JOB

Wanda Austin

I almost quit engineering before I found my dream job. As I was finishing master's degrees at the University of Pittsburgh, I interviewed with several companies and accepted an engineering job at Rockwell International in Anaheim, California. Things had begun going poorly there even before I started. Between the time I was hired and the time I started work, the company had a reduction in force, and many employees lost their jobs. The job environment at Rockwell was terrible. Workers were worried about being laid off, and they likely felt threatened by a young black woman with advanced degrees from the East Coast. I hated the job, and I began to reconsider my career choice— perhaps choosing engineering over applied mathematics had been a mistake.

When I was trying to decide what to do next, one of my colleagues told me not to judge a whole field based on a first job. I heeded his advice and ended up with a dream job at Aerospace Corporation. Working on systems engineering for satellites proved to be so much fun I would have done it even without pay. Rising through the ranks at Aerospace, I was eventually named president and CEO of the company. Now retired from leading a corporation, I have more time to spend with young people, to whom I provide guidance and mentoring and pass on lessons I have learned.

FAMILY VALUES

I was born in the Bronx near Yankee Stadium. My family lived in what was called "the Projects," a very low-income housing area. As a young African American girl growing up in the inner city in the 1960s, I was never told I could be a scientist or an engineer. My good fortune was having parents who valued education and knew that studying and hard work were keys to creating opportunities in life.

At first, I attended a local elementary school, but my mother thought I would get a better education in a different neighborhood. So I rode a bus across town to an elementary school in a white Jewish neighborhood, where I was one of only a couple of black students at the school. At my new school, in addition to learning from my classes, I also learned about different ethnicities and cultures.

My parents arranged educational trips and experiences for me. We visited Radio City Music Hall and Rockefeller Center to see plays, and we visited museums throughout the city. My Girl Scout troop made trips to the country. Our church also organized activities to get the kids outside the city. The whole community helped me access positive opportunities to take advantage of what New York City had to offer. These opportunities—to attend the best schools and to have a range of experiences—distinguished me from most of the kids I grew up with, who spent their entire youths within only a few blocks of where they lived.

My seventh-grade math teacher, Gary Cohen, recognized that I had above-average talent in mathematics. He gave me extra assignments and math puzzle books to read. Once when he returned a problem set that I had completed, he called me out in front of the class and said, "You are really good at this. Never let anyone tell you that you are not." Those words stuck with me; whenever I would start to feel discouraged, I would remember this message to aim high and that Gary Cohen believed I was capable of the highest level of achievement. I appreciated Mr. Cohen's vote of confidence, and I am still in touch with him today.

At the end of the seventh grade, I tested well in math and English. The school officials told my parents about a special program for high achievers to combine three years of junior high school into two. So a group of us skipped the eighth grade. Then in ninth grade I took and passed the admission tests to attend the Bronx School of Science. I started high school there when I was 13.

Wanda Austin grew up in New York City and graduated from the Bronx
School of Science in 1970.

A RIGOROUS HIGH SCHOOL

I found the Bronx School of Science to be a special place, with smart students and many resources. The students were very capable in math and science and able to solve hard problems. The facilities were amazing, with modern biology and chemistry labs like I had never seen in a school before. The school had computers long before the general public had heard about computers.

At Bronx Science, I took a full and rigorous academic program, including French and drafting, and of course also took advantage of its rich math and science offerings. But all through high school my social life was centered in my neighborhood. After school, I took three buses to get back to that neighborhood and my friends.

My experience in high school was not uniformly positive. There was one teacher who declared that I would never understand English literature and told me to just sit in the room and not cause any trouble. Those words discouraged me, but I was not one to go sit in the corner and keep my mouth shut. Bringing home bad grades was not an option in my home, so I pushed through that terrible English class and graduated as a well-rounded student with good grades. After that experience, though, I gravitated toward mathematics. In math, either I had the right answer, or I didn't have the right answer, but I never had a math teacher tell me that I was incapable of figuring it out.

Like most of the students graduating from Bronx Science, I planned to go to college; the only question was which one. I read guidebooks and found a college that had a math major and a good reputation and that offered me financial aid. The first time I saw Franklin and Marshall, a small college on a beautiful rural campus in Lancaster, Pennsylvania, was when my parents dropped me off there.

Out of 2,000 students on campus at Franklin and Marshall, only 20 were African American. Any member of a minority can struggle to find support networks that others take for granted. But as an African American woman from the inner city and a geek by anyone's standards, I could have felt isolated at college in multiple dimensions. I had to figure out how to cope. Mathematics turned out to be a refuge for me; the Math Department was small enough that everyone knew everyone, and math was a great equalizer. My math professors challenged me and expected me to perform just as everyone else did.

Overall, I found Franklin and Marshall to be a wonderful environment. The college was small enough that I never felt lost in a crowd of students. The professors were welcoming and would invite me over to their homes for dessert and to talk—what we would now call mentoring. Such visits weren't scheduled—they were spontaneous, just considered a regular part of life on campus.

During freshman year, I skipped my calculus lecture once. That afternoon I was at a coffee shop when Professor George Rosenstein saw me and said, "Hey, you missed my class this morning." Of course I had no excuse. He told me not to let it happen again. That level of caring and attention from the faculty was what allowed me to thrive in the college environment. Holding me accountable to show up to class wasn't special treatment at Franklin and Marshall. That was just how the professors engaged with all the students on campus. There was an expectation that each one of us would be part of the college community—as a participant and a contributor.

At Franklin and Marshall, I took my first class on feminism, I spent time at the Green Room Theater, and I developed a passion for watching

Shakespeare's plays. I immersed myself in the intellectual community that was so different from any community I had been a part of up to that point.

A SUPPORT NETWORK

As a math major, I took several courses from Professor Rosenstein and developed a relationship with his family. The Rosensteins invited me into their home: I would visit, have cookies and milk, and baby-sit their daughter. When I started to get serious with a boyfriend, I took him over to visit and get their seal of approval. My relationship with the Rosensteins was more than just about math or academics—they gave me the support I couldn't get from my family, who, without having had a college experience, couldn't understand the social and academic support I needed.

During my junior year, I went to England to study at the University of Lancaster. To be dropped into a different environment as an American in a foreign country was another horizon-broadening experience. Franklin and Marshall assigned a local family to look out for me so that I had a place to stay during breaks. The English educational system had 10 weeks of study and then four weeks of break. The four weeks gave me time to get out and travel—to explore Europe and see the world from a different vantage point. That exploring was a critical part of my education as I got to see our country through someone else's eyes. I realized how small the world is and how similar people's concerns are. The families I met were worrying about their kids' education, just like my family. They were also worrying about their freedoms, just like we were.

When I returned to Franklin and Marshall for my senior year, I thought about my possible career choices in math. Some careers would not require additional training—for example, teaching math or doing actuarial science. But I decided to continue on to graduate school in applied math, and George Rosenstein was instrumental in that decision. While I was in Europe on my junior year abroad, George had taken a sabbatical year at the University of Pittsburgh. When we returned to campus, he raved about the University of Pittsburgh and recommended the graduate program in applied math there, so that's where I went. I entered a master's program and studied applied math, including statistics and operations research.

Near the end of my first year at Pittsburgh, I realized that the undergraduate engineering students I was tutoring would go on to make much more money than I would. There was something wrong with that, I thought. So I went over to the engineering school to find out about its programs. There I met a professor who became excited about transforming me into an engineer. I worked with him on transportation systems, specifically the application of hydrodynamic fluid flow to vehicle traffic. We studied whether ramp metering improves or impedes traffic flow and how slow-downs propagate on the highway. It was an interesting real-world math problem. After two years at Pittsburgh, I earned two master's degrees—one in applied math and the other in systems engineering.

As I was finishing my studies at Pittsburgh, I started looking for a job. I interviewed with several companies, including Proctor & Gamble in Cincinnati and Rockwell International in Anaheim, California. I visited Rockwell in April, leaving my apartment in Pittsburg with ice on the windows and arriving at warm and sunny Anaheim, where I promptly went to the beach. So it was the weather, not the details of the job, that compelled me to pick Rockwell over Proctor & Gamble.

A TERRIBLE FIRST JOB

But my first job in California was horrible. Between April and August, Rockwell had laid off employees in the department I was supposed to join. The company still honored its employment offer and gave me a job. I worked on change detection in synthetic-aperture radar data, creating a computer program to see whether there had been any changes if a plane flew over a scene one day and then again the next day. It was an important real-world problem with many applications. But the mindset around the organization was negative. Employees were suspicious of seeing a qualified young person show up and learn their job skills. They were worried that I had been hired as a cost-cutting measure to replace them. Mainly fearful for their jobs, they were not at all welcoming to me. Another difficulty at Rockwell was the number of ways in which I was a minority at the company: I was an African American female in a male-dominated field, a mathematician doing engineering, an East Coast person suddenly transported to Southern California. Combined with the awful economic circumstances, these factors made for a very difficult situation.

After working at Rockwell a little while, I thought I might have picked the wrong career field. Based on a suggestion, though, I decided if I were to apply elsewhere, I might have a better experience. I interviewed at Hughes Aircraft and The Aerospace Corporation. Not hearing back from Aerospace, which was my first choice, I accepted the offer from Hughes. The week before I was supposed to start at Hughes, I heard that Aerospace was interested in hiring me. When I let them know I was just about to join Hughes, they promptly sent an offer letter, so at the last moment I switched companies.

A DREAM SECOND JOB

When I joined The Aerospace Corporation, I was introduced to the space business. From a systems-engineering perspective, I felt as if I had hit the lottery. I got to spend time at the secret US Air Force "Blue Cube" test center in Sunnyvale, learning about operational planning for satellite control. It was fascinating—I would have been happy to work there even if I weren't getting paid. I got to work out important problems at the leading edge of technology, such as detecting issues in orbit to prevent satellite loss.

I worked on satellite control and operational planning for about three years and then joined the customer-facing program office. Working side by side with an air force counterpart, I learned about the commercial side of the space business—how contracts work and how to structure incentives. Aerospace Corporation's mission is to provide independent technical advice to the US Air Force. Although my air force counterpart made the final decisions, I had a significant amount of responsibility, ensuring that our system worked properly within a huge system of systems.

After I had been working in the program office for two years, I attended a company cocktail party and talked with Eberhart Rechtin, the company's president. He told me that if I went back to school and got a PhD, many career opportunities would open up for me. I listened carefully and wanted to say back to him, "Getting a PhD really wasn't my plan. I'm married with two kids, and going back to school would be very, very hard." Instead, I followed his suggestion. Eb Rechtin subsequently was a champion for me through my career as well as a great mentor.

I looked for a flexible PhD program because I couldn't drop everything in my life and become a full-time student. I found such a program at the University of Southern California (USC), which was geared toward working adults. Classes were scheduled for late in the day, which allowed students to work half a day and then attend class.

I met with several professors to talk about research ideas and found a professor who was excited about working with me on system simulation. For my thesis, I worked on system dynamics and artificial intelligence with Professor Khosh Nevis. The goal was to create a natural-language processor—a tool that would allow a computer to parse a verbal description of a system problem and create a simulation model from spoken words. This was possible because you normally use just a few words to describe a typical system—the initial conditions, rates, and feedback relationships between elements. I created a program in which a user could speak directly to the computer and the computer would then accurately design and solve the simulation model. The program ran on a mainframe computer at USC, which I was able to monitor over the Internet from my desk at Aerospace.

> I have always worked long hours and often needed to take long trips, especially when there were satellite launches. There are periods when the company needs all hands on deck. If there was a meeting somewhere, I simply had to go. For independent launch reviews, I needed to be there to see the vehicle being transported from the C-130 into the building and prepped prior to launch. This added up to a lot of travel, which was unfortunate because I can't make up the time I missed with my family. But my job was not one that could be done remotely. I needed to be in the room with other engineers, working through hard problems.

At The Aerospace Corporation, most of the engineers had advanced degrees and about a quarter had PhDs. Getting a PhD put me in the category of engineers who proved we could do independent research and demonstrate critical thinking—engineers capable of formulating and solving hard problems. The company valued these skills because we were solving first-of-their-kind problems; there was no book on the shelf to look up the right answers to the problems we tackled. Our work spanned multiple engineering disciplines involving materials, batteries, sensors, and thermal issues, and each issue affected the system as a whole, including the ability to provide power, to fit inside a fairing envelope, and to be light enough to launch.

As a systems engineer, I came at a problem trying to understand these interactions—how everything fit together. I worked with electrical experts and mechanical experts as well as across disciplines. After seven years of working as a high-level systems engineer, I was appointed a principal director at Aerospace. This advanced leadership position required me to manage budgets, projects, people, and personalities.

BECOMING AN ENABLER

I remember the day I realized that I had moved from being an individual contributor to an enabler in that I was enabling others to solve problems. The team told me they had run into several barriers. Customer expectations were unclear; the team lacked access to required expertise; and they needed additional funding to conduct a critical lab test. I made some phone calls

Wanda Austin found her dream job as a systems engineer working on satellite control at The Aerospace Corporation. She rose through the ranks at the
company, serving as CEO from 2008 until she retired in 2016.

and contacted the customer. We renegotiated the schedule. I contacted a
colleague and arranged for the needed expertise. I brought the team back
into my office and told them what I had been able to do for them. I asked
them if that was enough to allow them to successfully complete the project.
The team members said yes, and I remember they were ecstatic.

But as the team left my office, I felt a sense of loss. They were working
on a great project and part of me still wanted to be a technical contributor
in the middle of it. I wanted to run calculations and predict system performance. But that wasn't my job anymore—my job was to ensure the team

Balancing the commitments of going to graduate school, working at The Aerospace Corporation, and taking care of two young children was very much a joint effort with my husband. I tell both men and women that when choosing a partner, you need to choose carefully. My husband recognized that I was not going to be standing at home with slippers and a cigar when he walked in the door at the end of the day. He was thrilled with that because he knew what I was working on and how much I enjoyed doing it. My husband was also an engineer—he worked on the shuttle program. So he understood how I could get wrapped up in the mission.

My husband was a fully engaged parent. When I went back to school for my PhD, we worked out schedules to make sure we would always have dinner together as a family. After dinner I would go to bed so I could get up early to study, and he would stay up to entertain the kids. On weekends, when there were birthday parties and soccer games, he handled those so I could have blocks of time for schoolwork. For me, being able to have kids, work, and go to school was very much about finding the right partner.

had everything they needed to succeed. It was a pivotal point in my career as I realized my value as a leader and enabler. Leading was important and something that I could do that they couldn't do for themselves. I felt a sense of reward, but it was very different from the rewards of being an individual contributor. After that point, I became more comfortable with my new role.

As a leader, I still needed to use the deep technical skills that I learned as an individual contributor. As president of Aerospace, I would conduct launch reviews of every satellite launch. Launch review is a grueling engineering exercise with many technical details. If I needed to recommend to an air force general not to launch due to some subtle engineering detail, I needed to convincingly explain our reasoning. I would have to be able to describe the problem, the risk to the launch, and the options for fixing the problem.

I have since retired from leading The Aerospace Corporation. I am working and traveling much less, and when I do travel, it is for fun. My husband and I took a trip around the world. We have four grandchildren, so we get to visit them. I wrote a book about leadership, and I speak about it to raise money for a science-technology-engineering-math student scholarship that we created. I still serve on the Defense Science Board and the NASA advisory council, and I'm a trustee to USC and a director for a corporation, so I am still keeping busy.

ABOUT THE SUBJECT

Wanda Austin is a systems engineer who retired in 2016 as president and CEO of The Aerospace Corporation. She has an undergraduate degree in applied math from Franklin and Marshall College, master's degrees in applied math and systems engineering from the University of Pittsburgh, and a PhD in systems engineering from the University of Southern California.

13 FROM OUTSIDER TO COLLABORATOR

David Galas

Shortly after I earned my PhD in physics, I decided it was time for a change.
I had entered college when the fields of
molecular biology and genetics were just
developing, barely a decade after the discovery of the double helix, but as I left
graduate school, these intertwined sciences were yielding more and more of
their secrets each year. I had always had
a dilettante's interest in the life sciences
but didn't know where to begin, so I
sought out as many distinguished biologists as I could find.

I met with James Watson, who had
helped discover the double helix; Walter
Gilbert, a physicist turned biologist who would go on to cofound Biogen
and Myriad Genetics; Harold Morowitz, a biophysicist from Yale; and Elbert
Branscomb, another physicist turned biologist at the Lawrence Livermore
National Laboratory. I found that even with my limited biological knowledge, in these conversations I could probe only two or three questions
deeply before coming to the response, "We don't know." But I also realized
that we now had the experimental tools to answer these questions.

For me, having fundamental questions everywhere and the means to
answer them is what defines the golden age of a science. So I decided I had
to participate in the golden age of molecular biology. I ended up doing so
as an academic scientist, government official, and entrepreneur. My career
has not unfolded as I would have predicted. For example, as a physicist in
graduate school, I would hardly have expected that my most cited paper
would be one I published as the chief scientist of a biotech company. At

every step of my career, I find that I have made the most progress by trying to see where the most exciting science is heading and by learning to collaborate with many others while using whatever insights I can bring to the life sciences from my training in physics and mathematics.

A ROVING CHILDHOOD

I was born in 1944 in St. Petersburg, Florida, where my father was finishing his training to fly B-17 bombers. He shortly thereafter left for England, from where he flew combat missions over Germany, but when the war was over and he was back from England, our military family moved around the country—to Illinois so Dad could go to graduate school, to New Mexico, back to England, and then to West Point, New York.

My dad was the first in his family to go to college. He was fascinated with planes and flying and had attended the US Military Academy at West Point, where he studied engineering. After the war, we first moved to Champaign, Illinois, where my dad studied for a master's degree in electrical engineering. We lived outside of town in rooms in a farmhouse, where I fed the chickens and pigs every day, followed Harold, the farmer, everywhere, rode on his tractor, and generally merged into the life of the farm. I remember shooting pigeons with Harold and taking them home to my city-raised mother for dinner. Though my parents were hardly farmers, I was raised on a farm for a few years, where the daily pulse of life was rooted in biology.

My interest in biology seeded on the farm grew from there. During my family's years in southern England, I went to school and lived most of the life of an English child of the early 1950s. I was exposed to and assimilated much of the culture of that time and place. Many of my friends enjoyed bird-watching and the now seemingly peculiar practice of collecting birds' eggs in the fields and hedgerows of England, activities that require immense patience and the acquisition of much knowledge of birds. After my father's cousin gave me a beautiful paperback edition of Roger Tory Peterson's illustrated guide to birds, which I largely committed to memory, I became completely fascinated by the ecology and habits of birds. Biology really began to penetrate my view of the world here, I think.

My father had accumulated a library of technical books over the years, which traveled around the world with us. Some of my earliest memories are of reading and barely comprehending science and engineering books from this library. At about age nine, I recall, I began really digging into my father's chemistry, calculus, and physics textbooks. At first, I understood only a little, but I understood more and more each year and was steadily drawn in. I read the calculus book through at about age 11 and did the problems without really understanding them. When my parents gave me a chemistry set when I was about 10, I nearly blew up my room. It may have been the hole in the ceiling that sealed the deal, but I decided then to become a scientist.

When we moved from England to West Point, my father was an instructor in the Physics Department. He became friends with the technician in charge of instructional demonstrations for the Chemistry Department. Seeing my fascination with chemistry, the technician arranged a wonderful opportunity for this 10-year-old kid. One afternoon a week I exited my school bus at the chemistry building. Mr. Rose would show me the current chemistry demonstration. I was immensely excited by these afternoons of synthesizing nylon, melting eutectic metal alloys in hot water, and so on—I remember many of these demonstrations to this day and still have the test tubes and other odds and ends he gave me.

In high school, my interest in biology waned quickly. The double helix hadn't yet made it into the curriculum, and biology seemed to be merely an exercise in classification and dissection, and although I liked to make drawings of biological objects, I found myself drawn rather toward physics and mathematics.

In the late 1950s and early 1960s—the Sputnik era—science education in the United States was buoyed by a national imperative to compete with the Soviet Union in science, aeronautics, space, and nuclear engineering. During my high school years, which I spent in Dayton, Ohio, there was also a big push for peaceful use of nuclear energy. I was motivated by the national sentiment, from watching the many kids' science shows on television, and from reading my father's textbooks. My interests in high school shifted toward physics, especially nuclear physics, as I understood it. Unfortunately, the science teaching at my high school in Dayton was not very good, and I learned physics mostly on my own. For a science fair project, I built an apparatus to show that the electron energy levels in the hydrogen atom are quantized. If you accelerate electrons through dilute hydrogen gas, at a voltage corresponding to an energy level, the current should drop there, and the hydrogen gas should radiate. The frequency of light should correspond to the voltage-accelerated electrons if energy were to be conserved, so this was what I wanted to show. I built the vacuum-tube apparatus for accelerating the electrons, and it somewhat worked. I then built a spectrometer with a cheap diffraction grating and tried to make this energy-conservation experiment work. It never really did, but the lure of physics had hooked me by then.

I entered the idea, design, and apparatus in a Dayton science fair, and it scored well enough that I got to go to a science fair in Chicago. That trip to the conference and the Chicago Museum of Science and Industry was truly inspiring, and the experience convinced me completely that I wanted to study physics.

MUDDLING INTO COLLEGE

My mother had never been to college, my father had gone to West Point and joined the military during the war, and my high school provided students with very little counseling or career advice. As a result, in 1962 I had no idea where I should to go to college or how to get in. During my senior year, I didn't even realize that I was supposed to apply.

The summer after I graduated from high school, my dad was assigned to Washington, DC, and I traveled with the family. I decided sometime in June or July that Georgetown University seemed like a good school, and perhaps I would go there. Visiting the college dean of at Georgetown University that summer, my father and I somehow appealed to him, told him that I wanted to study physics, and convinced him to let me enroll for the fall semester. I wish I remembered this kind man's name.

> There are passion-driven and peculiar creative projects you can do only in academia, there are massive endeavors you can undertake only with government backing, and there are rapid, focused, and applied team efforts you can best pursue in the context of a start-up company. I've learned that just as there's no single pathway from graduate student to successful scientist, there's no single pathway from basic research to technological innovation—in each sphere, different stages and methodologies inform each other in surprising and productive ways. At any one time, one needs to examine the problem and the available opportunities and make a considered judgment.

My first physics classes at Georgetown were a formative experience. My first professor, Dr. Beckel, was remarkably clear and very good at creative problems. He managed to engage my propensity for independent learning. He emphasized the value of the paradox in science and asked us to invent paradoxes related to what we were learning. I succeeded in finding one such paradox involving the difference between the forces on charges moving at different speeds, which requires relativity to explain it. I never forgot that experience or that problem, and Dr. Beckel loved it.

Studying physics at Georgetown was a great start, but I soon decided I wanted to attend a school that was stronger in the subject. I had heard that the two best places to study physics then were Princeton and the University of California (UC)–Berkeley. Our family couldn't afford Princeton, so after a bit of complex history I ended up at Berkeley.

Shortly after I arrived at Berkeley, I met Edward Teller, who was at the time a professor of physics and director of the Lawrence Livermore National Laboratory (LLNL). He recruited me for a job at LLNL that would shape much of my early career and allow me to pay for school. I ended up spending my time at Berkeley working three days a week as a programmer at LLNL and three days at the university.

In my senior year at Berkeley, my adviser pointed out that although I had taken many physics and mathematics courses, I wouldn't graduate unless I took some other course for at least a semester. Remembering how much I had enjoyed my birds, I signed up for zoology, which happened to be held on one of the weekdays I was in Berkeley. The professor was a wonderful lecturer, and I was amazed by how much biology had changed since high school. Notably, I learned about the double helix and what we now call developmental biology, and that field began to sink deeply into my mind. Although I wasn't immediately drawn back into biology, it had clearly become something different than it was before, a fertile and interesting field.

STAYING AT LLNL FOR GRADUATE SCHOOL

For graduate school, I briefly considered staying at Berkeley or moving south to the California Institute of Technology (Caltech). Berkeley's strength was in particle physics and the Bevatron on the hill, which was increasingly far from what interested me. Although I had greatly enjoyed the Physics Department with its eminent professors, some of whom were truly wonderful teachers, I also worried that I would get lost in such a large department. Caltech was smaller and had several professors with whom I wanted to work, but when I visited Pasadena in 1967, I found the smog so bad that I couldn't breathe. It could have been a bad day, but I made my decision and thought about staying at Berkeley.

Then a third option emerged. Edward Teller had recently started a department of applied science with students working at the national labs in Berkeley and Livermore but officially attached to UC Davis. Although the primary mission for LLNL was nuclear weapons, there was also amazing research being done in the labs, and there still is. One of my friends in grad school was using modifications of the same computer codes used for nuclear bomb simulation in reverse to simulate the processes of star formation. I was enthusiastic about the science being done at Livermore and was familiar with Berkeley and the Bay Area, so I decided to stay there for graduate school.

For my thesis, I worked with Harry Sahlin on the theory of excitations in superfluid liquid helium. Working in Livermore, however, meant that I didn't interact as much as I would have liked to with physicists at Berkeley or Caltech. But in some ways that suited my style. I was happy to bury myself in a problem and work on it until it was solved or understood. With my Hertz Fellowship, I was able to work very independently and spent a lot of time just wandering by myself in different fields of theoretical physics. I now realize that if I had interacted more with colleagues or taken specific courses, I could have learned much faster, solved problems more quickly, and tracked what was happening in the field much better.

My first publication was symptomatic of this lack of interaction. It wasn't related to my thesis; rather, it was a theoretical problem I had come across in my wanderings and solved by myself. My adviser played no role other than to provide general encouragement. I just told Harry that I would write it up, and he said fine. So my first published paper was a sole-author paper. I later learned that at my level of experience this was probably not the best way to do theoretical physics. Soon afterward I had a discussion with some physicists at Berkeley who published a paper on a related result about the same time. It was clear that it would have been much better to work together from the start, combine our ideas into a single paper, and dig deeper and go further together, perhaps opening an avenue of further understanding. That was an early lesson for me in the advantages of collaborating. I still needed to learn how to do it well, however.

My second paper was another small problem unrelated to my thesis that I worked on because Harry wanted both help and to help me, I think. I dug in deeper than I needed to for what he and Carl Jensen wanted to do, and I ended up finding a general solution to their problem on computing quantum-path integrals—so I ended up as first author. This was a different kind of collaboration, where my colleagues posed the problem and provided some initial results, and I pushed the problem further and did what I could do best. The process worked well and confirmed to me the value of collaborative effort.

MILITARY RESEARCH

To attend graduate school during the Vietnam War, I had received a draft deferment, so when I graduated in 1972, it was time for me to put in my military service. As a physicist with a PhD in the US Air Force, I would most likely have been assigned to one of the weapons labs in Albuquerque, New Mexico. But a colleague at Livermore arranged for me to get a job in the Vulnerability Task Force of the Department of Defense, an advisory group focused on the problems of protection of the US nuclear strategic forces. This assignment as a technical adviser to the chairman allowed me to keep my office in Livermore, have some time to work on my own research projects, and consider what to do next.

The projects in the Vulnerability Task Force were actually very interesting. For example, one of the projects was to determine if there were ways to spot submarines from aircraft or from space. Walter Munk, an oceanographer at UC San Diego, and I analyzed the possibility that submarines might excite enough internal waves in the ocean's thermal layers, coupled with the wind waves on the surface, to alter the waves' spectra in a way that satellites or planes could detect and spot subs. What we found could have upended a US strategic doctrine that depended on invisible, invulnerable nuclear missile submarines.

At the same time, I continued some of my theoretical PhD work on liquid helium and Bose-Einstein condensates, but I really wanted to branch out of this field and so looked for other interesting problems on which I could focus my energies. In the early 1970s, there seemed to me to be two areas really open to new ideas.

The first area was in astrophysics and cosmology. Pulsars had been recently discovered, and there were many new ideas about these massive objects billions of light-years away. Studying pulsars and other phenomena might allow us to look back in time at the evolution of our universe, with advances in computing power allowing us to do it all in simulations. But the distinct disadvantage of requiring access to big computers for cosmology research meant that there were only a few places in the world I could do this work at that time.

The other possibility emerged from my longtime interest in molecular biology. There were so many interesting problems to solve in this field in the early 1970s. For example, I had read about the lactose operon—a single genetic switch that responded to the presence of lactose and controlled the whole lactose-metabolism machinery. But there were many fundamental

David Galas (*left*) collaborated with Elbert Branscomb and Myron Goodman on protein and DNA studies in the late 1970s.

unanswered questions such as how the switch actually worked and if other such switches existed.

BECOMING A BIOLOGIST

At the time, I was interviewing applicants for the Hertz Foundation Fellowship and had the opportunity to travel. I used my free time in Boston to meet with and question biologists such as Wally Gilbert and Jim Watson at Harvard and Gene Stanley at Boston University about the field. In Livermore, I started dabbling in Elbert Branscomb's lab, talking with Elbert, who was very generous with his time and lab supplies, and doing bacterial genetics experiments when I could.

I learned with my hands and from my mistakes the truth of what both Gilbert and Watson had told me: molecular biology is quite different from physics. It's messier, the data are both sparse and complex, and even a theoretician needs to understand the details of what happens in the lab in generating data. To make progress, I couldn't just sit in an office pondering and teasing apart theoretical problems. I did some productive and new research with Elbert on the accuracy of information transfer in molecular biology but soon decided that I needed to dig deeper into the current state of the field in a broader way.

So with help and advice from Watson and through a longer story than I will relate here, in 1977 I found myself doing a postdoc at the University of Geneva in Switzerland and began furthering my effort to become a real molecular biologist.

After I returned from four years in Geneva, I took a faculty job in the Molecular Biology Department at the University of Southern California (USC). My research on transposable elements went quickly and well, and I received tenure within three years. In contrast to the more lonesome years doing physics, I realized I was now working very well with valued collaborators around the world. Perhaps because of that, the department made me chair in 1985, where I found my outsider's background as a physicist quite useful. Actually, I may have been made chair because I was young and didn't know any better.

The department chair is responsible for setting the graduate curricula, but making such curriculum choices is contentious among faculty even in

David Galas (*left*) collaborated with Jeff Miller on a variety of DNA research problems, including studies of the lac repressor, in the early 1980s.

mature fields. In molecular biology at USC, many faculty came from different fields, and all the professors seemed to want to teach material as they had learned it. I needed to mediate among the biochemists, the geneticists, and the researchers trained in molecular biology, each with his or her own list of essential courses for this new and ill-defined field. The only biology course I had ever taken was undergraduate zoology at Berkeley. Using my own education as an example, I was able to gently persuade the faculty that they should embrace the idea that they were creating a new field, not reproducing their own specialties, and that we needed to think about doing things in a new way.

THE HUMAN GENOME

In the late 1980s, the US Department of Energy (DOE) asked me to serve on a committee looking at how the department approached biological and environmental research. Our report recommended some radically new approaches to the program and that among the next steps of bringing new biology into each segment of the DOE's research should be the sequencing of the human genome, a project that the DOE had just begun. The DOE liked our report enough that they asked me to come to Washington and run its biological and environmental research. After refusing the offer a few times, I finally realized this was truly a rare opportunity. After all, it would happen only once that we sequence the human genome for the first time—so in 1990 I agreed to go to Washington on leave from USC.

By the mid-2000s, sequencing genomes would become nearly routine, but in the 1990s it was the kind of project that was simply impossible to do without a massive governmental investment. In my DOE job, our budget over the entire range of research areas was about $500 million per year, and I got to work with Jim Watson, who was directing the National Institutes of Health component of the project, until he finally left there in 1992 over disagreements with the director. This was an exciting time with a future waiting to be written.

It took more than three years to get our portion of the genome program working well, planned, and organized. When I left, the project was on track, but the actual genome sequencing had not yet really begun in earnest. An important, open question remained: How much medical impact the program would eventually have. I saw technical and commercial possibilities here. A biotech company might search through the genes that could be or were being sequenced, discover their biological role, and, it was hoped, have an impact on a particular medical need. With a number of other scientists and entrepreneurs, I cofounded Darwin Molecular Corporation to look for such genes and these impacts, and I ended up working in the company for more than five years.

I am very proud of the work Darwin Molecular did in the early days of research into the human genome. We discovered one of the first genetic variants for early-onset Alzheimer's disease, and even though this variant turned out not to be commercially interesting in leading to any product, it had a long-term influence on medical science. The paper describing this gene is one of my most cited papers, and there were a number of other significant discoveries. The scientific impact of this research, directed at medical applications, turned out to be significant and had a fundamental impact on science in immunology, the biology of bone formation, and neurodegenerative disease.

Since then, I have moved between academia and business—cofounding Keck Graduate Institute in Claremont, dedicated to training at the interface of biological science and commercial efforts, working at the Institute for Systems Biology and the Battelle Memorial Institute, and helping found five biotech companies. Whether working as head of a small research group, as an academic administrator, or as an entrepreneur, I have never lost my love of doing science and of trying to have an impact on the world.

It's been an interesting adventure, this journey through science, and the lessons I have learned are many. High among them are that surprises are frequent, collaborators can often open locked doors, and the wide fields of scientific knowledge are places of wonder that can also change the world for the better.

ABOUT THE SUBJECT

David Galas is principal scientist for the Pacific Northwest Research Institute and the chairman of the board for the Fannie & John Hertz Foundation. He received his undergraduate degree in physics from the University of California–Berkeley and his MS and PhD degrees in physics from the University of California–Davis and the Lawrence Livermore National Laboratory, where he was a Hertz Fellow.

14 TROUBLESHOOTER AND LEADER

Norman Augustine

When I was deciding on a career, I made a choice. I decided to study systems rather than the fine details of one technology. I first studied a complex technical system as a young aerospace engineer at Douglas Aircraft in the late 1950s. We were building Nike Zeus ballistic interceptor missiles, which flew at a supersonic speed of Mach 8.5 and acceleration up to 20G. The speed and stress were off the charts—far beyond anything previously attempted. It was a huge program, and we were suffering flight failures every week. I led a group in charge of failure analysis. To understand flight failures and recommend design changes, we needed to understand the entire system. That meant understanding how intense shock, heat, and vibration interacted with the aerodynamic structures, electronics, and controls.

This choice to study systems tends to be a one-way process—once you make it, it is very hard to go back. I once asked US senator Bill Frist (R–TN), who was a heart and lung transplant surgeon before he became a senator, whether he could return to a medical career. He laughed and said no, he had made a life-changing decision.

Studying systems is an extension of wanting to understand how things work, which is something I have been doing since I was very young. As a child, I liked to take apart mechanical toys and clocks and put them back together. I grew up as an only child in the mountains of Colorado and always liked math and science. I had wonderful parents who knew the importance of education and worked to inspire me. But no one in my family previously had the opportunity to go to college.

I attended East Denver High School, a large public school. Near the end of my junior year, a teacher I had never studied with, Justin Brierly, called me into his office and asked me what I was planning to do after high school. I said I wanted to be a forest ranger because I liked the outdoors. He yelled at me for not having sufficient ambition and handed me two applications to fill out. Then he threw me out of his office. I was offended but did as I was told. One was an application to Williams College and the other was to Princeton. My knowledge of the world east of Colorado mostly ended in Wichita, Kansas, where my uncle had a farm. I had heard of Princeton and knew it was east of Wichita, but that was about it.

I ran into some friends who had attended Princeton, and they convinced me that's where I wanted to go to school. I was eventually admitted to both Williams and Princeton, but Williams admitted me first. The cutoff date to

Norman Augustine grew up in Denver, Colorado, and always enjoyed math and science.

accept Williams' offer of admission was several days before Princeton was to announce whether I had been accepted there. That was a big problem because Williams required a $50 nonrefundable deposit. That was a lot of money, and I refused to send it in. My parents were aghast, but luckily for me I was admitted to Princeton.

FROM FORESTRY TO ENGINEERING

During a preliminary interview, a local Princeton alumnus had asked me what field I intended to study. I said forestry. He laughed at me, saying that forestry was not one of the subjects taught at Princeton. He said geological engineering was the major closest to forestry. So after arriving at Princeton, I studied geological engineering. However, I didn't enjoy it because I prefer subjects that are more exact.

Returning to Princeton from a date in New York City at the end of freshman year, I recognized a Princeton senior who had attended my high school. He had drunk far too much and was in danger of falling off the train. I pulled him to safety, and we struck up a conversation. He told me he was an aeronautical engineer, that aeronautics is a great field, and that I should study it, too. On the advice of a drunken schoolmate, I changed my major. I studied aeronautical engineering beginning my sophomore year. My courses focused on subsonic aerodynamics. I was particularly close to two professors—Courtland Perkins, who was chairman of the department, and David Hazen, my adviser. Perkins pioneered in-flight test analysis of aircraft stability and control and wrote a classic book on the topic that is still in print. Hazen was interested in subsonic aerodynamics and was an expert on wind tunnels.

During my senior year, I had a job at a research lab at the Princeton Forrestal campus. I was thrilled to be doing research and accepted the job without even asking how much it payed. I had been making about $2.00 an hour waiting tables, and the previous summer I had earned $1.69 an hour spreading tar on roofs. It turned out that my job as a research assistant paid only 50 cents an hour, but the pay didn't matter so much to me—it was great experience. That was the first of a number of occasions in my career when I accepted jobs that were not the highest paying but instead looked as if they would be the most interesting.

The encounter on the train that caused me to change my major was lucky, and my luck continued. Shortly after I graduated with an aeronautical engineering degree in 1957, the first Sputnik satellite was launched into orbit. Sputnik triggered a space race between the United States and the

Norman Augustine (*left*) studied aeronautical engineering at Princeton University and graduated in 1957.

Soviet Union and generated great demand for aeronautical engineers. So my timing was impeccable.

I continued at Princeton for a master's degree. For my research with David Hazen, I studied the aerodynamics of a double-slotted flap for vertical-takeoff aircraft. Again by luck, my thesis work came in handy years later. Just after I retired from Lockheed Martin, the US Air Force was considering canceling the Osprey Vertical Takeoff and Landing Program and so asked me to serve on a committee to investigate the program. Although the concept was very different from my thesis, I was familiar with vertical take-off and its associated control problems. Our group recommended that the engine be redesigned, rather than cancel the program.

After I completed the master's degree, I was running out of money and tired of studying. I also was more interested in building things than in doing research. I applied to a number of companies, including Douglas Aircraft, intending to work on airplanes. Through a set of circumstances that I never figured out, I was offered a job in Douglas's missiles and space group. I thought that Santa Monica, California, would be a nice place to live, so

After completing a master's degree in aeronautical engineering in 1959, Norman Augustine joined the missiles and space group at Douglas Aircraft.

I accepted the offer. I had been at Douglas about a year when I received a letter, forwarded through Princeton, offering me a job with the Douglas aircraft division. I told the aircraft division that I had previously applied to their division but was now working with the missiles and space division, liked it, and wanted to continue there. This was another example of how luck changed the course of my career.

At Douglas Aircraft, I was assigned to a research group in theoretical aerodynamics because it was related to research I had done for my master's thesis. I enjoyed the work but was interested in more applied areas. After a year or so, engineers working on the Nike Zeus missile program started asking me for advice on aerodynamics and what could have caused their flight failures.

LEADING A TEAM

When I was asked to lead a failure-analysis team, I jumped at the opportunity. I led a group of about a dozen engineers, none of us very experienced. We certainly weren't experienced at Mach 8.5 flight, but the issue was not our youth—in fact no engineer had experience in that extreme hypersonic regime. In development programs, you normally conduct a test flight every couple of months., We had the good fortune to fly on three test ranges almost once a week, which provided a great deal of data to analyze.

My advice to students choosing a field with opportunities: if your interests are compatible with going into areas that are new and burgeoning, you are always better off. But they are hard to spot. It would have also been hard for me when I was starting out. Today, genomics and cybersecurity would be good areas. Also advanced materials, informatics, and advanced semiconductors. If you get into a field that is really taking off, you won't have mentors, but you also won't be competing with someone who has 20 years of experience. One field that intrigues me is human–machine interactions, a field of artificial intelligence that will eventually allow talking to a computer just like one would talk to a person. Your computer companion will be able to understand you. It will require a lot better voice recognition than today. Driverless cars and robotics are other excellent areas just taking off, and a lot sooner than most of us realize or perhaps are prepared for.

Working with the failure-analysis team was probably the most interesting job I have ever had—engineering combined with detective work. The data were in the form of telemetry, movie video, and missile parts recovered from the White Sands range. We worked backward from the data to try to determine what could have caused the failure. Once we determined how to fix the problem, there would be a crash effort to implement the improvement and get back to flight. In one case, a pattern of shock waves had reflected off the nose of the bird, then off a fin, and then back to the nose. The wave then converted to concentrated heat and melted a hole in the side of the missile, looking just as if someone had used a blowtorch on it. It took a long time to figure out how that could happen. We were running wind-tunnel tests and simulations, but our computer models were very primitive at the time. You really had to understand the physics in order to interpret the data.

After a couple of years of leading the failure-analysis team, I was transferred to a group that was writing proposals. I found writing proposals very frustrating. We would write a proposal that I thought very good and we would lose. Then we would write one that I wasn't very proud of and it would win. I decided that if I were going to spend my career in the aerospace field, I had better be able to tell the difference between a good proposal and a bad one. I figured the best way to do so was to read proposals that other people had written.

I had some friends in Washington. I let them know that if an opportunity ever arose, I would consider a job in the government. At the time, Robert McNamara had just joined the Kennedy administration. He was looking for energetic young people to join the Defense Department, and I was offered a job working in the engineering section of the Office of the Secretary of Defense. The change would mean a big pay cut, but it was a fantastic opportunity. I was responsible for eight or ten programs, where I had an impact on funding and direction far beyond my age and experience level. It was a marvelous learning experience and very rewarding.

Norman Augustine (*center*) visited a defense contractor that manufactured helicopters when he was assistant secretary of the army in 1974.

The Office of Secretary of Defense was divided into two offices, one for tactical systems and the other for strategic systems. The tactical office concentrated on how to fight smaller wars, whereas the strategic systems office was about managing our conflict with the Soviet Union. My background was in ballistic missiles and space, so I initially worked in the strategic systems office.

MENTORS AND LEADERS

During my time at the Pentagon, I had several important mentors. One of them was Daniel Fink, the deputy director in charge of strategic and space systems, who was my boss's boss. I met with him fairly regularly to brief him on my projects. Dan was a wonderful, decent, and brilliant person who looked out for his employees. He had had polio as a teenager and was stooped and bent over, but that never slowed him down. He was very demanding and insisted that I get the job done correctly.

As the Office of the Defense Secretary grew to support the Vietnam War, the tactical group began looking for additional staff. They asked me if I was interested in transferring into their group, which would be a promotion. Looking out both for my best interests and that of the larger organization, Dan Fink suggested that I take the job even though losing my contributions would negatively affect his group.

Another mentor who helped me in my career was Melvin Laird, the secretary of defense when I worked at the Pentagon. Twenty years later he was a member of the Martin Marietta Board of Directors when the company was looking for a new CEO. I assumed that my boss at Martin Marietta would be picked for the position. There were also many others there who were qualified, with 20 or 30 years of experience, whereas I had been with the company for only a few years. Due in part to Laird's support for me, I was chosen to be CEO.

Most of the jobs in my career have required me to be a good engineer but also to have good management skills. I never took courses in management but sometimes wish I had. There was never a point when I decided to quit being an engineer and become a manager. I started out at Douglas Aircraft, and soon there were 14 people working for me. At Lockheed Martin, where I was CEO, we had 182,000 engineers and scientists employed by the company. Throughout my career, I have continued to take technical courses and try very hard to maintain my engineering skills to some degree. So I still describe myself as an engineer.

Like Dan Fink and Melvin Laird, most good leaders are focused on their mission and not on themselves. There's a tombstone in the British Officers' Cemetery at Normandy that has inscribed on it my favorite definition of leadership: leadership is wisdom, courage, and carelessness of self. To these three traits I would also add character.

I have had the pleasure of working with famous leaders such as General Omar Bradley and aviation pioneer Jimmy Doolittle, but also with others who are less known. These leaders had different styles. Some were inspiring speakers. Some, such as Dan Fink, were quiet. Dan could be abrupt, and he would mumble. Although the "packaging" was different among them, they all had certain qualities in common. They were able to motivate the people who worked for them to achieve their full potential. A sports analogy might make this point clearer: NFL coaches Tom Landry and Vince Lombardi were great leaders with completely different styles, but both possessed the qualities of wisdom, courage, character, and selflessness.

AN INDEPENDENT TROUBLESHOOTER

Since I retired, I have had the pleasure to work with some of the world's most outstanding scientists and engineers in many different fields. I have a reputation as a troubleshooter that goes back to my second job at Douglas Aircraft. When something goes wrong, I am often asked to look into it. And when there's a major decision to be made, the government often wants the decision reviewed by an independent group. So I am often asked to chair a commission that will last from months to years.

Some of the commissions have concerned aeronautical topics, such as the future of manned space flight. Others have been concerned with US leadership in education, science, and technology. Many of the commission topics have been outside of my area of expertise, but there's an advantage to knowing less about a subject than the committee experts. I ask lots of questions, and I'm not afraid to ask a dumb one. I once chaired a commission on particle physics, which was very hard. But my colleagues on the committee included Nobel Prize–winning physicists, and the opportunity to work with them was wonderful. More recently, I worked on a commission that examined the tax laws of the State of Maryland, and I am currently a member of a commission looking at the National Interstate Highway System, the first time the subject has been studied since the system was authorized in 1956.

ABOUT THE SUBJECT

Norman Augustine is an aerospace executive and has held a wide number of roles
in industry, government, and academia.
He retired as CEO and chairman of Lockheed Martin Corporation in 1997. He has
chaired many government commissions
on topics ranging from aeronautics to US
technological leadership and economic
competitiveness. Most recently, he has
been the chairman of the Review of United
States Human Space Flight Plans Committee. Augustine has BSE and MSE degrees in
aeronautical engineering from Princeton
University.

15 LEARNING BUSINESS BY FINDING A MARKET

Richard S. Post

I had the good fortune to grow up in Silicon Valley, where companies were constantly starting up or had already achieved scale. Engineers in the neighborhood were willing to answer my questions and get me parts to build my projects, which ranged from amateur rocketry to building a van de Graff accelerator.

Growing up in the Stanford University environment, where after World War II, Professor Frederick Terman (considered one of the founders of Silicon Valley) encouraged his students to start businesses, I came to believe that what one did in life was to get a PhD, teach, learn some great technology, and start a business. The spirit I grew up with still persists in the United States, and many readers will be as interested in starting a business as I was. It still seems to me to be part of our culture, but…how does one get started?

I believe the best way to learn business is just to *start* one—no course will teach you all of what you need to know.

The first company I cofounded was Applied Science and Technology, Inc., better known by its acronym, ASTeX. The other three founders and I had worked together either at the University of Wisconsin–Madison or at the MIT Plasma Fusion Center, where we had moved from Wisconsin to get more electric power for our electromagnets and to escape the public university purchasing department.

As a senior at the University of California—Berkeley in 1966, Richard Post built a small electrostatic accelerator out of surplus or salvaged parts.

A LARGE UNIVERSITY EXPERIMENT

At the Plasma Fusion Center, we spent from 1980 to 1989 building and running a magnetic-mirror machine to explore plasma confinement in a linear geometry. As the program neared its end, a group of us began to think about starting a business. Much of what we did in a large experiment gave us useful preparation for the future business venture. The MIT fusion project involved a staff of 40 but peaked at 100 with the inclusion of co-op students during the assembly of the machine. The sheer size of the project required a business manager to deal with the US Department of Energy (DOE) in reporting, planning, and budgeting. One of the engineers at the Fusion Center who wanted to focus more on business issues than on engineering joined the team and would become our CFO when the business eventually formed. To help us run the construction and operation of the experiment and deal with a large staff, our business manager and I started taking short courses in project and people management. For example, when we did our first "up to air"—breaking open the vacuum system to add diagnostics and make repairs—we found that it took too long, six weeks, to get running again. We started using our new project-management concepts and began preparing for the next up to air, running a load-leveling program. We asked all the scientists to generate a list of tasks and the help they would need to install the equipment. When we put in all their data, we found that all the work was to be done by the senior technicians, leaving little for anyone else. That was why the first up to air took so long. So we started looking at the work and at who could do what, moved the assignments around, and showed the reorganization to the staff. We had load-leveled the work. After objections were made and the plan was adjusted, we had forty 40-hour work weeks for the senior and junior techs, scientists, and students. The next up to air took three weeks, and, with practice, we eventually met our two-week target.

As far back as I can remember, I was interested in science and technology. In high school, I learned best by doing, not by watching. Senior year I was paid to work for five hours each day and learn for three hours. I got to play with equipment and build my own experiments. This practical, hands-on approach is the best way to learn how to be a scientist.

The second big challenge was budgeting, accounting, and control. MIT gave us a half-dozen accounts, and the data were always months behind the current known costs. Our business manager noticed that each account had an empty field, which we made into our subaccounts and which gave each scientist and engineer an account for each project. We stripped out the wages, so the amount they saw was just the amount they could spend. To keep them from overspending, we set up our own accounting based on requisitions. We had an engineer near retirement who was willing to be the gatekeeper and could understand the project and requisitions. Everyone submitted their requisitions to him. He okayed them; without his signature, purchasing would not act. MIT was a pretty freewheeling place at that time, where students could walk into a machine shop and start spending money.

Our budget ran between $5 and $10 million per year, so we got a lot of experience that we never had when doing smaller scale research. Figuring out how to manage this budget and all the employees was a great learning experience. Building the program from scratch, which included erecting a building to house the experiment, gave us experience working with companies that would build specialized equipment. Plus, we had to do a lot of hiring. Interviewing to hire so many people over a few years gave us the experience in hiring that we would later use in business. Although résumés were helpful, we used oral exams on the subjects the applicants should know because getting into the details of the work they did in their prior jobs was most valuable. I also noticed that the most critical people were the best judges in hiring new staff, whereas the more laid-back interviewers tended to judge unexceptional candidates as good enough. One big take-away from these interviews was figuring out that if an interviewee told us he or she was really good, but his or her boss was bad, we would be the person's next bad boss.

Our purchasing of custom-designed components and large vacuum chambers, electromagnets, and so on attracted large companies looking to get into fusion-reactor construction. We quickly learned, however, that big companies, which had the infrastructure to build jet aircraft, could not do projects at our scale cost effectively. Small companies with narrow specialization to match our needs were the ones that got our business. We would design modules for data acquisition, build a prototype, and then find a small company that would commercialize the design, sell us 50 units, make a manual, and sell the module to all the other fusion labs. For some of our microwave-power supplies, we wanted to use switching power-supply designs and so sought out companies that would be eager to work with us. By working closely with those companies, we learned a great deal about small companies and came to regard them as part of the project, managing them as if they were employees. The relations between us and these small companies were open, free, and based on trust.

Before we completed the research program, the fusion budget had peaked and was starting to decline. The linear-mirror program was losing out to the toroidal (tokomak) configuration, and I learned that we would be winding down. Some of my colleagues and I had talked informally about starting a business when the program reached an end, but in 1986 we decided it was time to get serious. We heard that plasma would be a billion-dollar market, and plasma physics and associated engineering is what we knew, so five of us—four from MIT and one from the University of Wisconsin—started having weekly Saturday-morning discussions on what kind of business to create. Our colleague from Wisconsin had developed a new approach to ion implantation, called PSII. Our lead technologist had been asked by a local semiconductor company to consult on a problem it was having with an oxygen ion source. He quickly realized that the problem was not the ion source, but the power supply that had been designed for industrial heating. The ripple was so large that it modulated the terminal voltage. On evenings and weekends, he designed the first switching power supply specifically designed to drive plasma processing with low ripple and good regulation. About the same time, the National Academy of Science issued a report comparing the state of semiconductor microwave plasma-processing development in the United States and Japan. It showed that multiple microwave plasma processes were already integrated in Japanese production tools, whereas almost none of the processes in the US equipment companies were integrated, and there were no US component suppliers. At this time, there was great concern that Japan would overtake the United States in the semiconductor industry, which led to our formation of Sematech, and, with lucky timing, we realized that we had something the semiconductor industry needed.

As the end of our research program at MIT became defined, it was time to start the business. To get ourselves ready, we attended the MIT forum, a weekly meeting where start-up companies presented their business plans to a panel, usually composed of a venture capitalist, an accountant, a successful start-up founder, and the audience. Our business manager and I took several short courses on starting a company, writing business plans, and so on. Perhaps the most valuable takeaway from this process was to "know your banker": you don't want to have to do this suddenly when you find yourself in trouble. Business relationships are based on trust, which takes time and interaction to develop. This relationship will pay dividends later. We started interviewing lawyers, bankers, and accountants, getting to know their expectations and gaining insight as to whom we would like to work with.

Richard Post points out details of a proposed plasma fusion reactor to MIT
colleagues Jay Kesner and Ronald Davidson in 1980.

AN EQUIPMENT SUPPLIER

What kind of a business would we like to start? Well, it would be much like
the other small businesses we had come to know as our suppliers at MIT:
we would build equipment and sell it. We decided to start with our own
money. We divided up the stock with no more than a two-to-one variation.
Each of us bought our stock, raising $150,000. We agreed to a five-year
vesting because we wanted commitment to build a company. After looking
closely at ion implantation for metal, which the Wisconsin program had
focused on, we found that the companies in this area were not doing well.
Although the results looked impressive, the market being developed was for
metal wear resistance for hip implants, punches, and so on, but companies
in this field were not growing fast, if at all, and one, Zymet, had already
failed. So we agreed that we would evaluate the process for two years. and if
we chose not to pursue the product, the company would have the right to
buy back the unvested shares.

We spent a bit of time coming up with a company name, eventually
arriving at Applied Science and Technology, Inc. The name put the stake
in the ground and signaled we were done with research and would now
apply our knowledge to products. Our business manager came up with a

great company's nickname, ASTeX. "TeX" came from a popular typesetting program at MIT, and it gave the logo a distinctive look, one that caught people's eye. One time I called a company to clarify an invoice, and the administrative person said, "Oh, you're the company with the little *e* in your name." As it turned out, small things like this were part of good marketing and essential in creating a brand.

Our technologist donated his power-supply design as our first product. The next question was, At what price should we sell it? We weren't competing with the microwave heating power-supply companies, so we felt free to sell at a much higher price. Our technologist had already sold two supplies to a local company as part of his consulting. We knew that the rule of thumb in the electronics industry was to sell at four times the bill of materials in order to cover material costs, development, manufacturing, overhead, and profit. As the bill of materials was about $2,500, a selling price of $10,000 per unit was suggested.

One major difference between leading groups at universities and leading teams at companies is how suggestions and criticism are perceived. In universities, there is competition of ideas, with comments about ideas somewhat separated from implications about the worth of the people. Employees at companies don't always see this distinction. I've seen employees take my comments very personally and interpret my comments on ideas as statements about which employees are good and which are bad. Working with teams in industry has taught me to be more careful with my words and more aware of the impact some comments can have.

My neighbor across the street had started a company selling replica mirrors for the medical instrumentation market. We would chat regularly, and he would give me advice based on his experience. He pointed out that one of the optical companies making lenses for satellites had developed the best lens available, and the marketing people asked the engineers what they could sell it for. They priced it to make a profit, but he said they could have priced it two times higher, and if they had, they would have been a great company instead of just a good company.

"Never let your engineers price your product," he told me. So, armed with that advice, I recommended that we sell the supply for $20,000 per unit. However, a complaint was that customers could open the box and think we were cheating them. We therefore spent several weeks discussing

Richard Post stands next to a partially completed section of the TARA
tandem-mirror plasma-confinement experiment at MIT in 1983.

the price and eventually agreed to sell each unit for $17,600. Some custom-
ers did complain about the price, but they bought it. After they got their
power supply, the complaints stopped, and they wanted help with the system design, which we were happy to give because we would in this way be
learning what our customers were doing.

IMPRESSING INITIAL CUSTOMERS

Our business manager got us a free ad in *Electronics Design* magazine, where
we offered the first microwave power supply specifically designed for
plasma processing. We got two critical customers immediately. One was a
visiting scientist at Penn State University, who was developing a chemical
vapor deposition (CVD) diamond reactor. The other was a research scientist
from Bell Northern Canada, who wanted to build an electron cyclotron

resonance (ECR) CVD system for processing telecommunication laser diodes. We asked these customers to visit us, and then we walked them around our experiment at MIT so they could see that we knew how to build hardware. As one customer said, "OK, I believe you guys can build anything. Let's go talk about what I want." We then went to our 600-square-foot company, a sublet off campus, to learn what the customers wanted. Because our first customers were research-and-development scientists, we understood their ways of thinking, and they understood ours. There was good synergy because we knew how to make plasmas and our customers knew semiconductors and material processing, both of which were new to us. This relationship allowed us to develop our business as we learned how to work with customers. However, we had a lesson coming as we moved beyond the customer and into the original-equipment manufacturer (OEM) market.

The Penn State customer wanted to make hydrogen plasma with a small amount of methane to grow synthetic diamond and asked if we could run our source in hydrogen. That evening our source was tested, and we got an order to build the first of our many CVD reactors. It immediately began producing high-quality diamond, and the word got around in the university community. The customers asked for new features, which, when added, generated new products and an expanded set of microwave components. Similarly, our first ECR source went into operation at Bell Northern and was used to make laser diodes at a time when the telecommunication industry was ramping up fiber-optic lines. This became our standard model: get the customer to say what he needed in addition to the power supply, and if the customer was a high-quality "lead" customer whose use of our product would influence others, we would develop other products to go with the power supply.

To go beyond these initial customers, we needed to get sales coverage and so began to look at hiring manufacturers' representatives. We had met a local rep who wanted to represent us throughout the United States, and we went to visit his company. As with all sales people, he made a great pitch. Although he didn't have reps over the whole country, he knew the best people and would hire them. On the way back home, we discussed whether we should take him up on his offer. All three of my partners wanted to take him up on his proposal, but I didn't get a good feeling about him and said no, that I would do the sales because we were not going to know our customer if we had this rep as an interface. It was a good decision; I learned a lot from the customers and enjoyed the interaction. We did, however, hire some local reps around the United States and the United Kingdom as well as a German rep with branch offices around Europe, but we found that direct sales, especially with the OEMs, was the most effective approach.

In our search for lawyers, we met with a small partnership that specialized in start-ups and then grew with them. Their experience and advice were essential in helping us at every stage of development. We first met with the partnership's founder, who gave us a summary of issues we would face and the services the law firm would provide. He introduced us to the partner with whom we would work. He then gave us some advice from his experience: "As your company progresses, there will be ownership issues. Someone might get divorced and have to split his stock or might want out and look for liquidity. This is going to lead to tension. Lots of founders get into nasty fights, which damage the business. If such an event happens, don't fight. Make a deal, give up the equity you need to, and move on, as your stock will be worth more by building a successful company. Focus on the company's success." We then got started with his partner and drew up the company documents, which included our individual contracts.

Working with our lawyer took several months. We had already invested our funds and were filing patents on a number of key technologies we had developed when friction started between two of the founders. Unfortunately, our contracts were yet not in place, and it soon became clear that the company was not going to succeed with both of these founders staying on, so we needed one to part—the job of telling him was left to me. I told him that we wanted to buy him out. But the fact came up that his name was on a key patent, and he would not sign it over to the company, as would have been required under our employment agreements. We didn't have a lot of money and didn't want to spend it on lawyers, but we got a good deal with a litigator to resolve the issue for a flat fee. This issue led to a meeting of all five of us, with lawyers to negotiate the deal. We agreed to pay $60,000 to this founder over the next two years, which would give us time to grow before paying. We were following our lawyer's advice, and we all shook hands—except for the person we asked to leave the company. It is not easy for everyone to make a deal, particularly when it involves money.

RAISING CAPITAL

After this difficulty was over and the company was growing, we were soon faced with the need for more cash. We finished the first year with revenue of $432,000 and a small profit, but we were growing too fast to generate the needed cash for operations. We visited the Bank of New England, which had a group that specialized in start-ups. We discussed our results and our cash needs and were informed that if we all wanted put up our homes as collateral, the bank would give us a credit line. As one of my partners told the bankers, "We believe in our business, but we want to keep our homes if we fail." We received some great advice in return: "You're a high-tech company. Go raise some equity and come back." How much? "$350,000." So the next question was how to raise equity. We decided to go to "friends and family" because we were already being told that we needed a businessperson to run the company. We were going into the next stage of our lives and weren't interested in being told what to do. We judged that if we grew the business profitably, we could gain the credibility to finance the company as needed. The CFO and I generated a business plan, which was more a vehicle to raise money than a plan for the company to follow. It summarized what we were doing and our guess as to how the products might sell, but it did not promise more than we were pretty sure we could do.

Making a business plan was something we did not know how to do well, but it included the goal to achieve revenue of $2.5 million in our second year and make a 10 percent pretax profit. To do this, we decided to raise $350,000, with an overallotment of $150,000 at a company valuation of $3 million. We all contacted our friends and were getting a good response. We discussed fund-raising with our lawyer and accountant and asked what to do if we have sufficient demand to sell the overallotment. "Take it," was the emphatic advice—we did. My brother-in-law, an investor in London, introduced us to Genevest, a European investment company out of Geneva. The three principals of Genevest came to review the business, and all three bought a unit for their own accounts. In 1989, a unit sold for $25,000, set at a "sweet spot" of what we thought our friends would risk and was the typical investment. Now, with the investment behind us, the bank came through with a credit line of $350,000, secured by receivables and a $100,000 equipment line.

Continuing to grow fast, we worked with Genevest to bring in $1 million in equity, and the head of that firm joined the board, along with a partner retired from the accounting firm we used. The board offered advice for our CFO on finance, and we continued to learn more about raising money through Genevest. After our first round of financing, I started to wonder how the investors had made their decision to invest in us. Although we were growing fast, I doubted that alone was enough to convince them. So on our second round of fund-raising, I asked the principals at Genevest why they had invested in the first round. I was told that two important factors in their decision were our enthusiasm and our customer list. We were selling to AT&T Bell Laboratories, IBM, Bell Northern, Penn State University, and Fraunhofer Institute—all organizations our investors knew. I was told that although they could not evaluate our technology, they assumed it must be good if those organizations were buying from a small start-up like ours. Many other factors went into their decision, too, but these were the consistent themes.

By offering stock in Europe at a time when there were not as many European start-up companies as there are today, we got premium pricing. We continued to develop other European banking relationships. My brother-in-law introduced me to American Equities out of London, where we could tap another group of investors. Just like getting to know your banker, you should get to know your investment bankers before you actually need them.

RUNNING A BUSINESS

During the initial start-up phase, I worked at the company only part-time, focusing on financing and contracts, because I had to wind down the MIT program. ASTeX had hired about 15 people from the MIT staff and helped to place all the other employees in MIT research programs or in local companies, so I was then free to work at ASTeX full-time. My coming into the business full-time produced some tension that I did not understand. The group already in place had a strong sense of ownership of what they had accomplished and may not have wanted to give up any of it. In fact, they had by now learned more than I knew about the day-to-day running of a business. I had a great deal to learn. We weren't doing research anymore. Operations management was now the key to success, but we didn't know much. We set up our production line and hired good assemblers, who were managed by one of our lead technicians from MIT. We added a purchasing manager who had run our local purchasing field office at MIT. He was very efficient and practical and added his knowledge to the growing company.

In our third year, we had a product set for oxygen downstream removal of photoresist (Ashing), ECR source for etch and deposition, CVD diamond sources, and all the components to allow the customer one-stop shopping. We had now learned enough process knowledge from customers, short courses, and reading that we agreed it was time to see if we could sell our products in the semiconductor market. I called the local office of Applied Materials, the largest of the semiconductor capital-equipment companies. Its senior technical managers were PhDs from Bell Labs, and when I met with them, they understood the physics of our various plasma sources. They gave me the name of an engineer who needed to build a microwave downstream Asher and told me to go see if I could help. I met with him, but the meeting didn't lead anywhere because he had his plan, and I didn't have the sales experience to break in. However, one of our customers at AT&T Bell Labs told the Etch Division vice president that he ought to come by and see what ASTeX was doing. He did, bought one of everything, and hired one of our Bell Lab customers to head his development effort. Unfortunately, that did not lead immediately to follow-on business.

We were anxious, but we knew that it simply takes time to get new technology into a company, and we needed our Bell Labs customer to get integrated. In a few months, we made contact and discussed the Asher program. Our customer worked with the engineer to put our plasma-production product on Bell Labs' wafer-processing module. He informed us that it met the company's performance requirements. However, he requested a number of changes to make the product fit on Bell Labs' tool, including longer cables so the power supply could be outside the clean room. These changes took up all the engineering resources we had.

By this time, we were located in an 11,000-square-foot building near MIT, and we had a good five-year lease with a clause that the landlord would have to provide all the space we needed as we grew, or we could get out of the lease. The landlord told me he had reviewed these terms with his lawyer, who had told him that all start-ups put this clause in the lease, but they never grow that fast, so go ahead and accept it. At the end of year three, we were at $4.3 million and about to outgrow our space. The landlord let us out of the lease because he could not offer us the 20,000 square feet we were looking for, so we moved out to Route 128 to take over half of what had been Digital Equipment's 60,000-square-foot remote-maintenance facility.

Now in our new facility, having increased our fixed costs and having finished a year at $5.6 million, we continued to work with our OEM customer. But this also reduced our ability to support new work for our traditional research customers and contributed to slower growth. It took about a year for our OEM customer to get his Asher developed and to start qualifying the process on wafers with his customers, and so there were no new orders. We didn't understand how long this qualification process would take, and tension began to build. The head of development questioned me on what we were doing and said, with a glare, "They better buy a lot!" Risk is not always enjoyable.

Finally, our customer called us in to negotiate a contract, and our purchasing manager, who was now head of operations, went out to negotiate. Part of our microwave components could be easily copied, so he took our patent on the product with him. We wrote into the contract that the customer could license this part of the product but had to acknowledge the

Richard Post and the staff of the TARA tandem-mirror plasma-confinement experiment celebrate initial operation in 1984.

patent in the contract because we would not have the funds to sue—nor would it be a great business step to sue our customer. Years later we learned that the customer was copying the component for a new product, so we showed the customer's purchasing manager the contract and evidence, and the customer paid the licensing fee.

DOWNSIZING TO STAY AFLOAT

While we waited nervously for orders on our new contract, trouble came again. The Berlin Wall was down, and Germany was restructuring, as was IBM. The telecom market was slowing down, so new systems from Bell Northern were delayed. Both the IBM and German research establishments, each of which had been about 10 percent of our business, stopped buying. Germany was sending money east. Our customers at IBM, who previously could buy whatever they wanted, were now without funds. We used to call IBM "Impulse Buying at $100k," but this was over, never to return. We were now looking at our first quarterly loss, with no certainty of when orders would pick up. We needed more money, and we needed to cut costs.

> Leading large groups and companies is really all about dealing with people and with personality types. Each department in a company—sales, marketing, and engineering—has its own personality that reflects the type of person drawn to the discipline. Much of the challenge of keeping a company running smoothly is making sure these different groups can work well together. When the different groups are unable to work together, then there are problems. The groups try to get deals done outside of the authorized channels created by the company. It's important to mix these groups up at times, such as when I assigned accountants to individual business units, where they could interact directly with the teams and better appreciate their own role in the company.

The CFO and I met. Both of us knew what had to be done and so had a list of cuts, which included a 20 percent staff reduction. We went over the situation with the managers to put the cuts in action. There was no resistance as the need was obvious. The staff cuts were made, and salary cuts were announced. We also gave new stock options as a retainer for the rest of the staff. However, even with these actions, we were going to run out of cash and were about to go out of covenant with the bank. Our CFO called our account manager and explained the situation. He told the account manager that we would raise new equity and that the downturn was market based, so we would recover. Because we had developed the bank's confidence prior to the downturn, the bank allowed us to continue to draw on our line, which allowed us to proceed with the financing.

I had been developing a relationship with another European investment bank, American Equities, meeting the managing director in London and having him over to see the company. We had agreed on terms to raise $2.5 million but had yet to set a date. Given the situation we were facing, I called the managing director, saying that we needed to raise money now. Summer was coming, and so, due to vacations, he suggested we wait to meet in the fall. I said we had to do it now or I would work with another bank (although I had none), so he agreed. We went to Europe to give investor lunches in London, Geneva, and Paris. We talked to customers of both American Equities and Genevest. The investors were nervous and challenged me to support the revenue projections. I put up new slides with customers' names, products, prices, purchase dates. That list calmed them, but they still questioned—How do we know? By this stage, the market and customers were better known; we were getting better information from our customers, including our OEM, as to when they would place orders, but this meant little to investors, who tend to want something tangible to build their confidence.

Shortly after we returned home, the lead banker called to say he was not sure he could get the job done. He had some indications, but not enough to close the order book. Then Genevest called to say the CERN pension fund was going to purchase $1 million of the deal. With knowledge of CERN's pedigree, all the investors bought, and the deal was completed. The advantage of having a member of your investment bank on your board of directors is that this person knows how well management is running the company and what the customers' issues are and can see the future better than any prospectus can explain. Get to know your bankers early on—you don't want try to get to know them when you are in trouble.

After seeing how powerful CVD diamond was and knowing how difficult it was to present a clear message to investors, we decided at investor meetings to give potential investors a piece of diamond one centimeter square and one millimeter thick and have them slice a piece of ice. The diamond cut the ice like a knife cutting butter. This allowed them to connect microwave plasma with the world of products that ASTeX enabled. I used to say, "We sell stock with diamond and make money with power supplies." If you sell tech stocks to investors, you need a way to generate an "aha, I get it" moment.

Another act of learning came from the layoff. As business recovered and pressure to get more hires came, I asked each of the managers if he would hire back the person he had laid off. Of the ten layoffs, only one was recommended for rehire. I then realized what had been going on: people had been hired to do the work that the most productive people thought they could give to someone else, so they hired lower-level assistants. By saying they would not hire the employee back, the managers were also saying they had made a mistake. With that knowledge, we adopted a policy of "all hires must raise the average." If we were going to beat larger, better-financed companies, our staff had to be smarter than theirs. Everyone knows that, but what standards will they use when hiring? As I had just seen, our managers were not hiring to raise the average, and the only way they could do that is if they hired someone smarter, more experienced, or quicker than they are. This is not easy for everyone to do. Some engineers had a hard time with hiring someone whom everyone in the hiring team thinks is better than he is. But once the concept was accepted by the company, peer pressure to do the best we can in every decision the company makes won the day.

Our OEM's product with our Asher was finally accepted by the semiconductor-production customers, and the orders began to flow in. It took about two years from the first sale to getting production orders. Six months was spent before the customer got his program moving, followed by a year of integrating and qualifying and another six months for the factory to qualify and for the semiconductor customers to qualify the chips produced on the new tool. The result was great all around, for our customer, its customers, and ASTeX. When our operations manager visited the customer's purchasing department, he was told, "Since you guys did such a good job with the Asher, take a look at ozone production. We have a serious problem." This led to our next very successful product, a semiconductor-quality ozone generator.

A NEW ORGANIZATION

We had been organized around sales, operation, engineering, and so on, but it was getting difficult to get a completely new product built, as opposed to an evolving product, because current customers would drive us to meet their needs. Sales was having a hard time influencing engineering to do what each customer told them it wanted in the product. Our revenue was about $10 million, but we also had a growing employee head count. Companies start to change as they pass 40 employees. To some of the early employees, the company was not the old ASTeX they knew. Some departments were viewing ASTeX as made up of "them and us." To paraphrase Max Weber, "all social organizations, if left alone, will seek to serve themselves." The more separated the department is from the customer who will not leave them alone, the more likely the department is to lose sight of why it is there.

To solve these problems, we organized the company along product lines: OEM, research customers and CVD diamond, sales/marketing, operations, and a new group, advanced technology, which focused on new products without the need to focus on daily customer issues. The employees were divided among these groups. Engineers had the biggest problem with this change because they were a cohesive team under the leadership of our founding technologist, and so some contemplated leaving the company. But the accountants felt new ownership—they weren't just overhead; they were part of the product team and heard reports from our customers. The old ASTeX vanished, and a new ASTeX emerged.

The product groups were now led by a product manager with product-line responsibility. The product manager, the sales person, and the engineers were aligned. The senior engineer was now going on sales calls, and although this might seem like a poor use of his time, it saved time, too, because he then knew exactly what the customers wanted. Previously, when the sales person came back with a customer complaint, to the engineer the complaint felt like a personal criticism. Now he was hearing directly from the customer, and so he behaved just like the sales person: he responded to the customer and was eager to make the requested change. Engineering was now in partnership with the customer. I asked an inside sales person what she thought of the new organization. "Great," she said. "Now I know who to ask when I have a product question."

Since the product manager was now a key person, one of the ambitious engineers felt the way to enhance his career was to become one. I told him he was already a great engineering manager, but with experience as a product manager he would become an even better engineering manager. He eventually became the OEM product manager. I later asked him, as one of the engineers who first disliked the reorganization, what he thought now. He said he would never go back to the old style of organization.

We started to talk to US investment bankers about going public. We were averaging a 50 percent compound-growth rate, and with the OEM business we could see coming, we needed access to capital, more than we were able to get from our European investors. At that time, in 1993, many companies were going public with $20 million in revenue. We wanted to raise enough money so we would not have to do fund-raising as frequently because sales would drop when so much effort was diverted to fund-raising. A small Wall Street firm then approached us, noting that it had successfully taken out other local high-tech firms. Our board was in favor of going public, so we did it the year after completing a $9 million year and establishing a plan for $13 million in the coming year. We got a valuation of $45 million and raised $15 million. We had the balance sheet we needed to grow the company, and it enabled us to grow to $80 million in revenue before needing more capital.

The operations manager now ran the OEM group, so we brought in a new operations manager who had managed operations at a Polaroid plant. It was time to bring in more skills. We were running about $10 million per year, and the OEMs expected better documentation of manufacturing. Sematech was offering a course in Total Quality Management, which we took advantage of. One of the sales admins came back from the course saying, "We thought we were screwed up, but you should have heard about these other companies!" Educating our employees was more than having them learn specific skills—it was way to help them gain a better sense of business through their counterparts in other companies. Applied Materials' supplier days were a big help because they brought in speakers to teach suppliers how to run more efficiently. Applied Materials needed to lower its costs every year, so helping its suppliers get more efficient improved its own supplier base.

The talk that made the biggest impact on us was on lean manufacturing by James P. Womack, author of *Toyota, the Machine That Changed the World* and *Lean Thinking*. It was my first look at the company we needed to become. Management training in organizational behavior and the individual was important to get a better understanding of management dynamics and its impact on the organization and everyone within it. Our CFO was a self-taught accountant, so an accountant board member recommended that the CFO take the Harvard Business School Executive Education course on finance and control. Based on the CFO's positive experience in this course and my realization that I really did not know how to run a business, as opposed to making and selling products, I took Harvard's nine-week owner/president management program. Learning how a lumberyard operates and deals with problems through the case-study method was absolutely essential preparation for managing the high-tech business as we grew.

The advanced-technology group developed the world's best semiconductor ozone generator used in the TEOS/ozone deposition of silicon-dioxide films in semiconductor processing. The ozone generated had a low level of impurity and was very stable. We showed the generator to Applied Materials and were first told we were too late, but that they would look at it. We shipped our first unit, and a few days later Applied Materials ordered three. We were about to win the key component for a $1 billion product line. We had reorganized just in time. The new operations manager was put to work preparing for our first clean-room manufacturing.

The product manager reported back from the contract negotiations that the customer wanted to pay only 20 percent more than our bill-of-materials cost, but the volume would be big, much larger than our Ashing business. We decided to take the business and planned to work on cost reduction. A cross-functional team from assemblers to engineers was put together to come up with steps to reduce material costs. The first action had to be to remove the engineers from the team because they kept wanting to redesign the product. They instead became advisers to approve changes. With production of several hundred units per year and with suppliers providing input as to what we could do to reduce their cost, two years later the bill of materials was cut in half, which improved our profit margin to the degree we needed. One of the most productive members was our loading-dock manager. When everyone gets to participate in the company, making key contributions to the team, it is amazing what you can get from the employees.

We naturally had to expand our customer base and worked hard to develop new semiconductor OEMs. Our customer for the Ashing product kept having problems with our components. Suppliers are rated on quality, so in this case it didn't matter if the problem was generated by the customer's assembler. One service tech would be there every week. Our customer was growing so fast that as soon as we had train one group of assemblers, we would find when a failure occurred that those assemblers had moved on, and the new group who replaced them had no idea what the components were, and, moreover, the instruction manuals would be gone. We opened a local field office to support our business, but we wanted to make it easier for our customer to install our product and eliminate production-line issues. Since the trend in the industry was to have the supplier provide more of its product as a subsystem to flow into their lean-manufacturing lines, we conceived the "top-lid solution."

All the components would be integrated into one box that could be installed directly on the customer's process chamber. Because the unit would be installed in the clean room, a module-status light was used to identify which subsystem was not working, thus enabling repair in the field by modular replacement. Our engineers had to put on bunny suits with gloves and change all the modules, as our customers would have to do. We showed the concept to the two major US capital-equipment suppliers, our current customer, and that customer's competitor. Our current customer said it would stay with what it had, which it perceived to be less expensive but never factoring in the cost incurred in manufacturing. The potential new customer's marketing group wanted to buy the component box, but its head of engineering would only allow his own solution. I worked with our market manager to poll the customers on their preference for their Ashing source. The key customer told the new customer that if the new customer did not offer our product, it would be much harder to buy the latter's tool. There was a showdown meeting between the new customer's marketing group, armed with that assessment, and its engineer. The marketing group won, and we had an $11 million contract.

A LEGAL DISPUTE WITH A CUSTOMER

Good news, you would think, but our major customer's product group claimed that we "stole" its technology. This customer had worked with us so long and our product was so imbedded in its tool that its engineers, who were never there in the initial development, could think that. Although this was not at all true, the customer's product group said it would get a new supplier for the new 300-millimeter tool set. It even paid more to get a Japanese power supply to feed the plasma source the customer had licensed from ASTeX. As the customer's competitor started winning market share, some individuals in the product group blamed us, and the head of the division believed them. He lost stature for losing market share, but this was the price we would have to pay to not be a one-time OEM player. We showed the product group that we had offered this product to our current customer first, and it had turned down the offer. That approach did not solve anything, though; no one gets anywhere by blaming the vendor. We then asked the division head to review the product group's claims of invention. We sent the customer's legal department all our documentation and patents. The legal department reviewer said there was no evidence for the customer's claim, but the relationship was severely damaged. This problem was not something we were prepared for or understood how to effectively deal with, but perhaps it just could not be avoided. We learned a great deal about what might happen but not how to manage it.

I taught in the Peace Corps in South America from 1966 to 1968 before starting graduate school. My job was to teach physics to science teachers at the National University in Bogota, Columbia. I think that taking time off before graduate school can be very helpful for students in figuring out what they really want to do. Going to graduate school just because you have "the grades to get in" is not the right reason. The Peace Corps was an opportunity for me to learn a bit more about myself and to grow up a bit. It gave me a broader perspective on the state of education internationally, and, by happy coincidence, it was also where I met my wife, Janet.

Although the company continued to grow at 50 percent per year, we were not as profitable as we needed to be to get full valuation for our stock. We spent heavily on research and development, which was delivering the products, and cutting back on these efforts seemed the only way to improve profitability. Nevertheless, the question remained: How do we drive this cutback through the company? Our comptroller made very detailed quarterly profit-and-loss statements for each product line, so not only did we set goals focused on revenue growth, but the product groups were now thinking they had a responsibility to work with manufacturing to get the profit up. We realized we had to train all our key managers, sales people, and engineers on how to run their profit-and-loss accounting. There were about 50 people in the training program, which started with basic accounting. We hired a teacher to spend several days going over the basics, and then our comptroller taught the details of our profit-and-loss system. The product groups started working with manufacturing to understand their costs and immediately found ways to reduce product cost. The company had new profit-margin targets and lots of incentive as all the company employees had stock options, which began increasing in value based on profitability.

The training program for our employees did several things. We had placed importance on their development and given them the power to execute. They started to understand their own relationship with investors and what was needed to get recognition. We then added a two-week general-business program. We had hired a marketing manager, a retired Harvard MBA, for marketing and to provide some "adult advice" to all of the managers. He worked with a Harvard professor he knew to put together a course covering all the standard MBA topics. The course was great, and the employees loved it. The company was transforming from a product-development company into a business. Profits were rising, and the stock price was lifting.

The problem in managing the semiconductor-equipment space was the cyclicality of the business. There were many equipment companies and many factories, known in our industry as "fabs." When Microsoft would come out with a new operating system or Intel with a new process, new chips manufacturing would be needed. All fabs would buy at once or lose market share. Typically, in two years the fabs were up to capacity, and there was an oversupply. The OEMs would use service spares, return tools, and so on to conserve their cash, so the supplier base business would be more cyclical than our customers' business. We got very good at sensing a downturn and cutting costs quickly.

Richard Post and MIT colleagues Donald Smith and John Tarrh cofounded Applied Science and Technology Inc. (ASTeX) in 1987. ASTeX, now a division of MKS Instruments, manufactures reactive gas products for semiconductor processing.

MARKET DIVERSIFICATION

Due to the cyclical nature of the semiconductor business, we had purchased two companies, one in laser power supplies and one in radio-frequency power supplies for magnetic resonance imaging and semiconductor markets. In the downturn in 1998, business dropped for four quarters, with management cutting further each quarter. At the bottom, half of our business was from laser and medical products, and we were glad to have diversified. Then came the orders. The next quarter the book-to-bill ratio was 1.7, meaning high demand, and we had to build up manufacturing again. The nature of this market makes outsourcing essential, but you want your suppliers to be diversified. When there is a layoff, knowledge leaves, too. When business returns, new hires will need to be trained to get the quality back. In the meantime, quality suffers. If your company outsources to a manufacturer that has only 30 percent of its business in semiconductors, then having a 50 percent drop in your business is only a 15 percent drop in the manufacturer's.

The company recovered from the downturn and finished 2000 at $139 million, making a record $1.09 per share. We had an offer to buy the company. After 13 years, the founding managers wanted to move on. The cyclicality had taken an emotional toll. In spite of difficulties with some managers at our largest customer, our sales to that customer had continued to grow it and now constituted 55 percent of our business. Although our having a lead position at the strongest semiconductor company was impressive, being so dependent on a single customer made the investors nervous. Our components could not be easily replaced by competing products because of the high cost of requalification, but investors feared that we would lose pricing power on new products. If we merged with the acquirer, business from our large customer would drop to 30 percent of the combined business, and the cost of supporting fabs in Asia, where the majority of the new business was moving, would be reduced. We could solve many problems by selling, and so we did. ASTeX has continued as a brand and became a successful division of MKS Instruments.

Not only ASTeX grew, but its employees did as well. The founder who had left the company founded his own successful company. Our office manager, who was a Jill-of-all-trades, no longer found ASTeX exciting because her work focus became narrower and more specialized and so left it to found her own company, which she then sold and started another. She credits what she learned from our start-up with enabling her to do the same. Both of our first two operations managers started their own businesses. One became our information-technology manager, but since he knew operations, he developed all the sales-account-management software. This software became shared by several companies founded by ASTeX employees, making the product better for all. Another of the founders left to start his own company, which was also purchased by MKS Instruments. He subsequently left that company and started yet another. I went on to found NEXX Systems to make the whole production tool, targeting the need for packaging semiconductors for cell phones. We developed a new tool set, waiting for cell phone manufacturers to put a computer in a phone—finally, the iPhone—and NEXX's success.

What makes me most proud, though, is the benefits our employees realized. We paid college tuition, which enabled several of our technicians to complete their engineering degrees. One of our assemblers, who had been born in Vietnam and whose family was forced back to China after the war, decided to go to America. Entering Hong Kong, she was caught and put in a camp for two years, after which she was sponsored by Quakers to come to America. She took electronics courses in night school while working in a garment factory. When we needed our first part-time assembler, we offered her the job, which she took, with the condition that we get a microwave so she could heat her lunch. A microwave was rescued from my neighbor's garbage and repaired, and so she started. We needed her to work full-time right away, so she quit her other job, joined us, and got her first stock option. I wasn't sure if it would be easy to hire people to work for a small company of a half-dozen people, so I asked her why she took the job. "Whenever I needed a tool, you guys bought it for me," she said. "The other company issued tools and would not buy anything else." Since employees are our biggest cost, almost anything that makes them more efficient is free. In our 10-year celebration, we asked two employees who had purchased their homes with their stock options to come up. One was our first assembler. A year later, as that employee continued to work for MKS, her daughter won a full scholarship to Harvard. Only in America would that be possible.

ABOUT THE AUTHOR

Richard S. Post is a retired serial entrepreneur and scientist who has recently become a serious amateur astronomer. He has an undergraduate degree in physics from the University of California at Berkeley and a PhD in physics from Columbia University.

THE ACADEMICS

16 THE RIGHT ENVIRONMENT

Jessica Seeliger

After six years of graduate work in Steve Boxer's biophysical chemistry lab at Stanford, I was feeling burned out, and I considered leaving the academic career path. I interviewed for positions in industry and management consulting. But when I met business consultants, I kept comparing them to professors like Steve, and they were never as smart or as excited by their work. I also missed the openly critical atmosphere that is central to science and found I had different values from people I met in the business world. These factors helped me realize that the best direction for me was to stay in an academic setting and continue on my planned path to become a professor.

I am now an assistant professor in the Medical School Department of Pharmacological Sciences at Stony Brook University. My husband, Markus, is also a professor in the department. The environment in which I work is very important to me. During my years of scientific training, I experienced a range of working environments: from overly competitive, joyless places where researchers toiled around the clock to kinder, joyful labs led by supportive, nurturing group leaders.

In my research lab at Stony Brook, I strive to create a positive learning environment that incorporates aspects of the labs where I had my best experiences. I want my lab to be a place where respect for colleagues is an important value and that allows a good work–life balance. I want the graduate students and postdocs in my lab to feel they can and should have a life outside of their research jobs.

AN INTEREST IN NATURE

I grew up in Oberlin, Ohio, a small town near Cleveland where Oberlin College is located. Both my parents were physicians who had come from Taiwan to finish their medical training. In elementary and middle school, I showed an aptitude in math and science and was fascinated by structures in the natural world, such as seashells, trees, and leaves.

In high school, I was drawn to biology and chemistry. Junior year I had an inspiring high school chemistry teacher, Mrs. Carlton, who took note of my interest and ability. One day when Mrs. Carlton was sick, she asked me to substitute for her and teach the class. I really enjoyed this first teaching experience, although some of the students may not have appreciated having a peer for a teacher. Junior year I wanted to learn more chemistry, but our small school didn't offer advanced-placement courses. So I studied the material on my own and enrolled in a thermodynamics course at Oberlin College with Professor Norman Craig.

In the spring of senior year, I asked Professor Craig if I could work as an intern in his lab for the summer. I had done well in his class, so he agreed. Professor Craig trusted me, a high school student, to work independently in his lab, even alone on weekends. To assist with his research on the spectroscopy of halogenated small molecule, I learned to operate and maintain a high-vacuum system, make measurements on a range of analytical instruments, and interpret the data on my own. It was an amazing introduction to research and a seminal learning experience for me.

I attended college at Harvard and majored in chemistry. My work in Professor Craig's lab led me to concentrate on physical chemistry rather than biochemistry. I wanted to go to Harvard in part because both my older brother and sister had gone there—by the time I applied, I had visited my siblings there a number of times. In addition to studying chemistry, I also fulfilled premed requirements, in part based on guidance from my parents, who recommended that I become a physician because it would be a good, secure career, although I did not end up applying to medical school.

Each time I changed labs, I have changed research areas—from protein folding to spectroscopy of reaction centers to tuberculosis. In making the decision of which project to work on, I have mainly relied on my gut feeling more than careful analysis. And it has always worked out.

At Harvard, in an environment that I didn't enjoy, I encountered a number of researchers I wouldn't want to emulate. Lab mates competed with one another, and some researchers seemed only to be looking out for themselves. There were fights over instrument time and in some cases a lack of respect for proper safety practices. Our lab was working at the forefront of a hot research area, and there was plenty of good science being done, but I also witnessed results published that I knew were not reproducible, which made me feel disillusioned. Although my mentors were attentive and inspiring, my overall research experience at Harvard made me question whether a career in academic research was for me.

Professor Norman Craig trusted Jessica Seeliger to work independently in his lab at Oberlin College when Jessica was a high school student. That early research experience set her on a path to become an experimental scientist.

While at Harvard, I also was a tutor. I enjoyed helping students understand subjects that I found exciting. Based on how much I enjoyed teaching, I decided that I would still enjoy an academic career as a research professor. However, I did not find any role models whose work–life choices made me feel that I could follow the same path.

For a time, I was one of the leaders of Harvard's Women in Science organization. As part of that organization, I interviewed Dr. Marcetta Darensbourg, a visiting chemistry professor from Texas A&M. I told her about my interest in an academic career but that I didn't like the work–life balance of the Harvard professors I was observing. I had other interests, such as playing the viola, and wasn't sure I wanted a career that would require such single-minded pursuit. In response, Professor Darensbourg told me about her passion for horse riding and working with horses. She said that she was still dedicating a significant part of her life to all her passions, of which science was just one.

Professor Darensbourg's story about horses and the possibility of having a balanced life as an academic had a strong and lasting impact on my thinking. Years later, when interviewing for a faculty position at Texas A&M, I visited Professor Darensbourg and recounted the story she had told me about her interest in horses. I told her that her matter-of-fact acceptance of herself as a multifaceted person and successful scientist had been an inspiration. She was surprised that our short encounter had so influenced me, but she was extremely pleased that I had taken her advice and pursued my dream of an academic career.

MOVING TOWARD BIOLOGY

Although I studied primarily chemistry in college, I have always been interested in both chemistry and biology. For my graduate research, I decided to work on biological systems in the interdisciplinary field of biophysical chemistry. I applied to Stanford and was accepted into its PhD program, but I deferred for a year to do a master's degree in biology at Cambridge University. My research there was in the lab of Alan Fersht, a distinguished chemist who has done fundamental work in protein folding. I really enjoyed my year in England; it was an opportunity to study abroad, to learn a lot of biology, and to experience a positive working environment.

Fersht's research lab was located in an academic center funded by the UK Medical Research Council. In addition to graduate students and postdocs, the center also employed full-time research staff. Staff members, who were generally older and more experienced than the postdocs, were very helpful.

Whenever I asked, these staff members would pause their work and spend up to a whole day to train me on a piece of equipment or teach me a new protocol. Such a positive and welcoming environment seemed like a revelation compared to what I had experienced at Harvard. I was also impressed by the "tea culture" in England. Instead of working heedlessly around the clock, as I had seen some researchers do at Harvard, staff scientists in Fersht's lab would arrive in the morning and go home to their families in the evening. They would take both midmorning and midafternoon tea breaks. Although staff scientists in Cambridge worked fewer hours than their US counterparts, they completed their experiments by careful planning and by making efficient use of their time.

Another highlight of my year in England was meeting my future husband. Both Markus and I were doing research on protein folding, so we really understood each other's work. When I returned to the United States

Jessica Seeliger (*standing, fourth from left*) did her graduate work in the lab of Professor Steve Boxer (*standing, far right*) at Stanford, finishing in 2006. She worked independently in Professor Boxer's lab, which exposed her to a range of research topics but, she felt, also made her less productive than if she had worked on a single well-defined project.

to start graduate school, Markus stayed in Cambridge to finish his PhD. We maintained a long-distance relationship for 18 months until Markus took a postdoc at Berkeley.

I started graduate school at Stanford with the intent to join a spectroscopy lab but decided instead to join the group headed by Professor Steven Boxer. Compared to researchers in other labs I had worked in, the researchers in Steve's group had a deep understanding of both biology and physical methods for investigating biological systems. I chose the Boxer lab because of the training I would receive in the group rather than because of a strong interest in any a particular project.

Steve was a hands-off adviser. Though he closely monitored whatever results I produced, he gave me a great deal of independence to choose the direction of my research. The disadvantage of this approach was that I tended to wander between projects. I was not as productive in graduate school as I might have been had Steve and I agreed on a well-defined project. But I learned about many more research areas and gained more practical experience during my time in the Boxer lab than I would have had I been more focused on a single topic.

A FAVORITE PROTEIN

I worked on several topics for my thesis, all studying my favorite protein in the whole world, the reaction center produced by purple photosynthetic bacteria. My main topic was using ultrafast spectroscopy to understand light-induced electron transfer in the protein. A second, related topic was labeling the protein with spectroscopic probes. By introducing a reporter at a specific spot in the protein, I could measure the local electric field to better understand the electron-transfer process. I also worked briefly on a third topic—supported lipid bilayers as models for biological membranes and the diffusive behaviors in those membranes.

I love the photosynthetic reaction center. It is almost 100 percent efficient at turning light photons into the movement of electrons, creating energy from light far more efficiently than any solar cell created by man. Some researchers have tried to re-create reaction centers using light-absorbing dyes that are similar to those found in these proteins, mimicking nature to make an efficient solar cell. But these efforts have yet to match nature's best work in photosynthesis.

My PhD research took six years to complete, and by the end I was feeling a bit burned out. Unsure whether I wanted to stay in academia, I explored industry and management-consulting jobs and applied for postdoc positions.

When interviewing at consulting firms, I realized that I was interested mainly in the problem-solving aspect of management consulting rather than the financial and other aspects. Ultimately, the process of looking at jobs in the commercial sector helped me realize that I still wanted an academic career. I wasn't sure if I could become a professor of the same caliber as Steve Boxer, but I decided to try. I chose to do a postdoc at Berkeley for two reasons—it would prepare me for an academic position, and Markus was doing his postdoc there.

As was the case for where I would work on my PhD, my choice of lab for the postdoc was more about environment than research topic. When I interviewed in Dr. Carolyn Bertozzi's lab, everyone I spoke to was energetic and truly excited about science, about what they were doing, and about working in the lab. The opportunity to learn from colleagues like them was as appealing as the topics I could work on. Once I joined, I found that many lab members were also interested in academic research careers. They were highly supportive and engaged and would attend practice talks, read research proposals, and provide insightful feedback. Scientifically and socially, it was the environment I needed to support my career goals.

For my postdoc research, I studied the microbe that causes tuberculosis. My postdoc experience in Carolyn Bertozzi's lab at Berkeley reaffirmed my interest to stay in academia. I found Carolyn to be a positive influence. Whereas Steve's uncompromising rigor had honed my critical scientific acumen, Carolyn's style of encouragement was just what I needed after my graduate school experience.

By 2008, Markus had had a productive postdoc and was ready to apply for a faculty position. At that time, I had been in my postdoc for only a year and was not yet ready to do the same. He applied to several institutions, including Stony Brook University, which he knew to be proactive in recruiting academic couples. The possibility that I might be able to get a position at Stony Brook was an important reason that Markus chose to accept a position there over other offers he received. It was a calculated risk, which ended up paying off.

I applied for a faculty position at Stony Brook in 2009 and received an offer. Markus and I now have identical tenure-track positions in the same department. We are very lucky. I am one of very few of my friends who solved the "two-body" problem such that both members of a couple have academic jobs in the same city. Making that happen required some planning, but mostly a great deal of luck.

Having enjoyed both my experience tutoring at Harvard and my graduate school duties as a teaching assistant, I learned that I prefer working

Jessica Seeliger and her husband, Markus, waited until 2017 to have children, when they had positions in academia. She finds that having a child has been a surprisingly joyful and a welcome contrast from her work.

directly with individual students over teaching classes. This is the type of teaching I do now—one-on-one research mentoring of the graduate students and postdocs in my lab. Though I occasionally lecture, in my Medical School department the teaching duties for junior tenure-track professors are very light, so I haven't yet been responsible for an entire course.

BALANCING ACADEMICS AND FAMILY

My current work is on fundamental processes in the tuberculosis bacterium—how its cell wall and membranes are made. I'm interested in the physical processes of how the lipids are moved around en route to the exterior of the bacterium, where they have both structural and biological roles. Those roles have implications for how the bacterium makes people sick.

As an assistant professor, I will eventually go up for tenure, although there is no strict clock on when that will happen. The funding environment has been quite hard lately, and that is one of the criteria that favors tenure. I have been supported by smaller grants but so far do not have the major funding that is traditionally expected of a professor being considered for promotion. Markus does have tenure, and it is good that one of us crossed that bar before we had our first child. It means 50 percent less stress in the household.

My son, Carl, is now one year old. I chose to wait fairly long before having a child. I was always ambivalent about having children because I recognized the challenges of being an academic while raising a family. From the beginning of my career, the family issue was on my mind, and as I got older, I concluded it was now or never. I wanted to feel ready but realized that I would never feel completely prepared. As it turned out, the adjustment was easier than I thought it would be. One reason is that Markus and I live close to campus and have been able to split many childcare duties—I'm grateful for his support and the flexibility of our jobs.

I wasn't planning to be a superwoman role model who would go back to work immediately after having a child. But within a week after giving birth to Carl, I was feeling well enough to attend group meetings. Being able to continue working helped me maintain my sense of self as a scientist; I had been afraid that the role of mother might overshadow everything else I had been creating. I appreciate being able to retain my dual roles of scientist and mother.

> There are people who are grand-problem driven, but I was driven by excitement over science. I was always very open to tackling any problem if I was excited by the methods being used and the different approaches or the avenues opened up by pursuing a particular problem. I have not been motivated by a grand vision but by what I might learn. My career has been a bit like a random walk through science. It's been a lot of fun, and I've learned a lot. I've been fortunate enough to be accepted into great labs and to learn from experts in the field.

Nevertheless, I underestimated how much having Carl would affect my productivity, especially in the first six months. But having a child has also helped me prioritize my tasks. I realized that some things really don't have to be done, or at least not right away. For example, I no longer reply to email as quickly as I used to and have found that the sky hasn't fallen. Yet.

Having a child has been surprisingly joyful and a welcome contrast from my work. I was worried that it would be hard to give up time from my lab and teaching to care for a child and that being a mother would cause me to lose my focus. Also, I have never really spent much time with small children, so I was not sure what I would do with a young child. But it's true that children reintroduce to us the concept of play, something that gets beaten out of us by the time we are adults. So it has been nice to reacquaint myself with simple pleasures. Carl is happy for no particular reason—he is happy because he exists. And that is a reminder for me of what life can be like.

ABOUT THE SUBJECT

Jessica C. Seeliger is an assistant professor in the Department of Pharmacological Sciences at Stony Brook University and runs a lab focusing on membrane biosynthesis, structure, and behavior in bacterial pathogenesis. She holds an AB in chemistry from Harvard University, an MPhil in chemistry as a Churchill Scholar at the University of Cambridge, and a PhD in biophysical chemistry from Stanford University, where she was a Hertz Fellow.

17 THE ADVANTAGE OF NOT FITTING IN

Stephon Alexander

My career has been a challenge in figuring out how to navigate two worlds—I never fit the stereotype of either a theoretical physicist or a jazz musician. It was during my first postdoc in London that jazz started to inform my physics. The Imperial College in London was the center of string theory in Europe, and I was studying there, trying to understand the fabric of space and time and hoping to discover something new about the cosmology of the early universe. Sometimes I would bring my physics papers to the jazz club, and sometimes I would bring my  saxophone. For a while, I kept the two worlds separate, but then I started to see connections between the music and the physics, and musical analysis began to inform my research.

I am now a professor of physics at Brown University, directing a research group working at the interface of cosmology, particle physics, and quantum gravity. I also advocate for historically underrepresented groups in the sciences, including those who, like me, are from economically poor backgrounds and the first in their family to go to college. I am proudly both a physicist and a jazz musician, and visitors to Providence, Rhode Island, are welcome to patronize the local restaurant where my friends and I perform our jazz improvisations.

AN IMMIGRANT IN NEW YORK CITY

I was born in Trinidad and moved with my family to New York City when I was eight years old. I attended public schools in the Bronx, where, during my early years, I was just an average student. Looking around my neighborhood, I felt that no matter how well I did in school, my future was constrained. My life was inevitably going to be dismal since some of the young guys who were my friends were mostly ending up in jail or other bad situations. My parents valued education, but from my viewpoint it didn't seem to matter how well I did in school.

I attended the John Philip Sousa Junior High School and enjoyed playing basketball, reading comics, and playing the saxophone. One day in eighth grade we had a special guest at an assembly. An older man in an orange jumpsuit brought a boom box onto the stage and introduced himself as Fredrick Gregory, an African American astronaut. After we enjoyed hearing the music, the astronaut explained why he was visiting. He said that although it's powerful to own a radio and hear the beat, the real power is the ability to make a radio. That knowledge had helped him become an astronaut. Gregory had learned about science, gone to college, and earned an engineering degree. His presentation was compelling, and that day was the first time I thought about studying science.

Near the end of my year in eighth grade, my music teacher told me that I was one of the best music students he had taught. He offered to help me get into the High School of Performing Arts. It was a tremendous opportunity, but I had other ideas. I was more interested in science than music and chose to attend DeWitt Clinton High School, a regular public high school, and the second largest in New York City.

At DeWitt Clinton, I still was not a great student. I would cut classes that I found boring and take a bus to join my friends on the basketball court. We would take breaks to rap and to break-dance on flattened refrigerator cardboard boxes. Some of the other kids who were cutting class believed in aliens from outer space, and others had odd spiritual practices. We all wanted to escape from the dismal prospects the future held for us.

ROLE MODELS

During sophomore year at DeWitt Clinton High School, I met a teacher who became my role model and would make a huge difference in my life. Daniel Kaplan, my high school physics teacher, was teaching us because of his passion for education. He had a PhD in physics from Princeton, had worked on radar during the Korean War, and was also an accomplished musician. He could have had many possible careers. But what he really wanted to do was teach high school students.

One day in class Mr. Kaplan took a tennis ball out of his pocket, threw it in the air, and caught it. He asked the class what the velocity of the ball was when it returned back to his hand. The class was silent, but Kaplan noticed my attention. I told him that when the ball returned, it would have the same velocity as when it left his hand. Mr. Kaplan beamed a big smile and told us about a principle of nature called the conservation of energy.

After that physics class, I started seeing the world in a different way. Growing up in the Bronx, I was not a big fan of reality. Kaplan's teaching made me realize that reality is not necessarily what it seems. Physics became for me the gateway to unveiling the secrets behind the reality; space and time were not what they seemed, and physics became my escape. There could be a path in life where I could accomplish something.

My math teacher, Daniel Feder, was also an important influence. I studied with Feder all four years of high school. He was doing research in number theory while teaching high school math. He walked to school every day, through the rain and the snow. Both Kaplan and Feder modeled for me what it meant to be a dedicated teacher.

I hadn't considered the possibility that I could go to college. I grew up in a neighborhood with a lot of immigrants, and college was not something we knew about. When Mr. Kaplan suggested that I might go to college, I told him that my family couldn't afford college. He said there were scholarships for talented students like me, and he would write a letter of recommendation. I thought that was a fine idea. My family was also thrilled—they didn't know I might be eligible for a scholarship.

Once I demonstrated that it was possible to go to college on a scholarship, I became a family and neighborhood role model. In the years after I graduated, all my brothers and many of the kids in our neighborhood began to go to college on scholarships.

PASSIONATE ABOUT PHYSICS

At the end of high school, I became really passionate about physics, reading any physics magazine or book I could get my hands on. When I applied to colleges, the admissions officers heard that loud and clear. I knew I wanted to go to a highly ranked, competitive college and applied to about 14 of them. From my experience at a large high school with 6,000 students, I knew I would prefer the individualized attention and mentoring available at one of the smaller colleges.

During the college application process, I met a woman named Susan Sharin, whose father had attended my high school and had a college scholarship named in his honor. Susan took me under her wing and became my college adviser. She suggested Haverford College as a possible fit and took a group of us from Dewitt Clinton on a weekend visit to Haverford. During that visit, I met Professor Jerry Gollub, a famous physicist who later became another important adviser. Gollub interrupted our Saturday tour to ask how many of us were interested in majoring in physics. Gollub took the two interested students to his lab to show us a chaos experiment in fluid dynamics. Having a world-renowned scientist come to campus on a Saturday to greet me and show me his lab convinced me that I wanted to go to Haverford.

Haverford was academically far more challenging than high school. In high school, I was the top dog, but at Haverford the other students seemed like superstars. Not being one of the better students challenged my confidence, but I was helped by having really good professors who cared. Another way I coped was through sports. Although I had never done any sports in high school, at Haverford I walked onto the track and cross-country team, which helped me focus and prioritize my time.

> I wrote one of the first dissertations on string cosmology, working at the interface of two fields, each of which was relatively well developed. That gave me an opportunity to say something new, which opened up doors for me. That is where the new discoveries will happen. In terms of a career, especially if the student wants an academic career, working in a new field or between existing fields can lead to all sorts of research opportunities.

I was a passionate and enthusiastic college student, but my grades were not very good, especially during freshman and sophomore years. In junior year, something clicked, and my grades improved. Then that summer I had a pivotal experience doing research at Carnegie Mellon University. I worked in Professor Mark Kryder's lab using high-power short-pulse lasers to study the motion of magnetic domains in magneto-optic hard drives. We applied magnetic fields to drive the domains and measured their hysteresis behavior, imaging the domains in real time. I saw that research is different from course work. I decided that for my career I wanted to get paid to do research, and that meant I needed to go to graduate school.

In addition to taking physics at Haverford, I also took most of its sociology courses. Studying sociology was a way to understand the forces that have shaped society and to understand my own situation—what it meant to be a young man of color in the world. In addition to classic texts by Hagel, Durkheim, and Weber, I also read works by Ira Reed, the first black professor at Haverford and a prominent sociologist. Through sociology, I gained a new language and accessed literature and research on my social reality. Haverford is a liberal arts college, and I took that label seriously; studying the liberal art of sociology was a way of liberating myself.

One of my physics professors, Lyle Roelofs, believed in me and encouraged me to go to graduate school. He is a computational theorist, and I did research in his lab during both my freshman and sophomore years. He got to see my research aptitude—even though I did not have the best grades in the classroom, he believed that I could be a good research scientist. Professor Roelofs contacted the Brown Physics Department and successfully advocated for me to be accepted there.

BECOMING A THEORIST

At Brown, I initially planned to become an experimental physicist. I joined a lab doing laser experiments, but the professor was very hard on me when I made a mistake. It turned out that I wasn't actually that good in the lab—I was better suited to be a theorist. I was attracted to the power and economy of equations. For example, I loved being able to explain a wide range of phenomena just with Maxwell's equations. With a simple set of four equations, I could explain what happens in the sun or in a far-away galaxy or in the lab downstairs. It made me feel like a wizard—with my pen I could manipulate these equations and in doing so virtually manipulate physical worlds. The feeling I got by working with equations reminded me of the feeling of escape from reality I experienced when as a high school kid in the Bronx I first discovered physics. I also realized at Brown that there are things about physics that we don't know and that there was the possibility that I could author my own equations and predict new phenomena. Even though that level of success seemed unlikely, I decided it was a game worth pursuing and a risk worth taking.

I worked with three professors at Brown. My first adviser was Leon Cooper, a renowned physicist who won the Nobel Prize in 1972 for explaining the quantum mechanical basis of superconductivity. Cooper loved working on interesting and seemingly insurmountable problems, regardless of discipline. When I joined his group, he was trying to construct a theory of memory based on neural networks and quantum mechanical models. My work in Cooper's lab was on so-called unsupervised neural networks, which can train themselves to learn new memories. By using quantum mechanical analogs in neuroscience, I saw the value and beauty of applying patterns from one field to another.

My second adviser was Mike Kosterlitz, also a famous physicist who won the Nobel Prize in 2016 for solid-state phase transitions and topological insulators. I finished my thesis in the new field of string cosmology with Robert Brandenburger.

Just as at Haverford, compared to the other graduate students at Brown, I again was not one of the best, not one of the top dogs. But my advisers saw my enthusiasm for physics—they saw my spark. They got behind me and encouraged me, and for that I am so grateful. As a graduate student, I really needed that affirmation, and if they hadn't taken an interest in me and been so encouraging, I don't think I would have stayed in graduate school.

My thesis topic was investigations at the interface of string theory and cosmology. Brandenberger and I were wondering if it were possible that the constants of nature, such as the fine structure constant or the speed of light, vary in time. I showed that string theory predicts that the constants of nature do vary in time. That was an unexpected result and attracted quite a bit of attention.

After finishing at Brown, I had postdocs at the Imperial College of London and Stanford. At Imperial College, I continued my work combining string theory with cosmology to say something about the early universe. I was interested in how galaxies, stars, and planets came into being out of a chaotic, featureless, early universe and whether string theory could say something about those structures.

During my postdocs, I wasn't thinking about choosing topics to further my career and possibly obtain an academic position. My research projects were topics I was interested in, not driven by what others might find interesting. I wasn't thinking about competition in the field; only later did I notice how competitive the other researchers were, angling for prestige and credit in order to be considered for job opportunities.

The path I took was actually dangerous because I was working on hard problems and could have gotten stuck without any results to show for my time. Then I wouldn't have been able to get another job. But I got lucky. I was able to write a paper that helped start a new subfield. I was able to provide the first model of string theory for cosmic inflation.

The way I do research is to evaluate a wide variety of ideas and try them out without prejudice or bias as to whether they might be true. Rather than starting with equations, I try to have mental images in my head to represent the physics. I will imagine something that I do understand and then extend them by analogies to map into the unknown. I like to use music analogies because I am so familiar with them.

It was challenging at first to be a string theorist bringing something as unusual as musical analogies to physics. Sociologically already atypical in the theoretical physics world, I was hesitant to go down a path that would contribute even more to that perception of me. At the same time, I was getting more involved in the local jazz scene—first in London and then at Stanford. For a long while, I kept the two worlds separate in my social interactions. But I eventually felt as if I were being a hypocrite and came out of the closet, so to speak. I made it clear to the world that I was both a jazz musician and a theoretical physicist.

At a postdoc at the Stanford National Accelerator Laboratory, I moved from string theory into particle cosmology, which is the relationship between high-energy particles and cosmology. There are questions that both fields of physics try to address, such as why the universe contains more matter than antimatter even though you would expect an equal amount.

During my summers, I did a lot of volunteer teaching at a program run by Jim Gates, the American string theorist. I like teaching because it is a way for me to revisit areas that I need to understand better. After two post-docs and based on my teaching experience, I decided I wanted to become a professor. Over a period of two years, I applied for many jobs and had a number of interviews but didn't receive any offers. I thought that I might have hit a glass ceiling and that getting a faculty job was not in the cards for me.

A DIFFICULT CHOICE

Then I got an offer from Penn State, which has a great Physics Department. I really wanted to stay in the Bay Area and was head-over-heels in love with someone there, but I made a sacrifice to get a first faculty position. It meant the end of my relationship, which was very hard. In hindsight, I chose my passion for being a physicist over other important parts of my life. At the time when I moved from beautiful San Francisco, on top of the hill, to the middle of Pennsylvania for a faculty job, I really questioned whether I was making the right choice.

At Penn State, there was a lot of pressure to get funding. Being a professor was different from what I had expected; there were many job duties besides research and teaching. I was lucky to get support from other faculty members, especially Abhay Ashtekar, director of the Gravitational Physics Institute. For my research topics, I started to pursue more traditional areas but also spent time working on wild, crazy ideas. By working in both areas, I would be able to show, when it came time to apply for tenure, my ability to do traditional research as well as to work on risky ideas.

In jazz, when you are refining your improvisational technique, it is best not to think too closely about what you are playing—you are exploring. In the middle of an improvisation, you might play a note that is the wrong note or a chord that is the wrong chord. And the music flows from how to work with that note or that chord and make it sound right. Similarly, there have been times when my research seemed to be heading in the wrong direction, but in the spirit of improvisation I was able to take that turn and make it into something useful.

For example, at Penn State I spent time talking to both the cosmology theorists on my floor and the condensed-matter theorists downstairs. At the time, it was not obvious what connections condensed-matter theory had with cosmology. In hindsight, 10 years later, I can see that the papers that resulted from those collaborations have become topical. People are noting ways that tools from condensed-matter physics can be used in cosmology—for example, to understand dark matter. I am proud to have written some of the initial papers connecting these fields.

My approach has always been that of an outsider. It is related to sociology, my background growing up, and my place in society. I am reminded of what Groucho Marx asked: "Why would I want to be in a club that would accept me as a member?" For me, the idea that success is enhanced by being in the club was not true. My stance was not so much about defiance but about the realization that when you fit in, you might have to worry about staying in the club. And there are penalties for going elsewhere, which feels threatening to the other club members. I eventually became comfortable being the outsider. And since I was never an insider, I had no problems being the cosmologist who took a risk by talking to a condensed-matter theorist; I didn't have to worry that colleagues might laugh at me for an unlikely approach, and many times that approach actually led to new understandings. For me, there has been a real advantage to being an outsider and not fitting in.

ABOUT THE SUBJECT

Stephon Alexander is a professor of physics at Brown University and the director of the Brown Presidential Scholars program, which provides scholarships and mentoring to promising students in science, technology, engineering, and math fields who come from disadvantaged backgrounds. He has an undergraduate degree from Haverford and a PhD from Brown University, both in physics.

18 THE LONG WAY ROUND

Kathleen Fisher

A special day during my sophomore year of college had a huge impact on my life: it gave me a career direction, and on that same day I met my first husband Steve.

I was taking a computer science course, and in a discussion after class the instructor told me about a fundamental result in the field called the halting problem. Alan Turing solved this problem in 1936, proving that you cannot write a program to determine if an arbitrary program will get stuck in an infinite loop or eventually halt. The result is deep, but the proof is simple, and once you understand it, you can write it down in five minutes on the back of a napkin.

After seeing the halting problem, I was hooked on computer science—and after that class finished, I began dating the instructor. By the time I went to graduate school three years later, Steve and I were married. A year later we had a daughter, Elaine.

My career since then has been a careful dance, honoring commitments both to my family and to the field I have come to love. Sometimes—as when our young daughter kept me grounded during graduate school—these commitments were synergistic, helping me to remain focused and effective. Other times, such as when my career and my ex-husband's career were geographically limited by our commitment to live near our daughter, one commitment constrained the other.

That day when I learned about the halting problem set me on a 25-year winding path to achieve a longtime goal: to be a professor of computer science and to work on research problems with important applications, ranging from programming languages to machine learning to cybersecurity.

CHOOSING A DIRECTION

I was born in San Marino, a small suburban community in Southern California near Pasadena. My dad was an investment banker, and my mother was a history major who was also trained in engineering. She worked in engineering for years and provided an example for me that women could do math, science, or anything else we put our mind to. That was the attitude of my parents, who were supportive of all the things I wanted to do.

San Marino High School was a high-powered public school, with plenty of advanced-placement and honors classes. I did well but was not drawn in any particular direction, unsure whether I would major in the sciences or the humanities in college. As a high school senior in the mid-1980s, I didn't really know what the field of computer science was. I only knew that one of the students who had gone to my high school was majoring in it in college, and that it seemed like an odd choice.

> Every career has a big-picture idea. Doctors heal people; academics teach students or make discoveries. Each also has a mundane, day-to-day experience attached to it. It's the day-to-day that influences whether you enjoy it or not, and the big picture almost doesn't matter if you don't enjoy the day-to-day. From the beginning, I chose to do the things that I enjoyed doing—programming, then programming languages. The big-picture idea didn't have to align exactly with what I wanted all the time because I enjoyed what I was doing every step of the way.

If I had a favorite subject in high school, it was English. I'm still close friends with my sophomore-year English teacher, Joyce Steece, who taught me how to write. One memorable assignment I recall was to interview people in three different jobs and write a report about them. Since I was editor of the yearbook, I chose a magazine editor for one of the jobs. I was also captain of the women's golf team in high school, so I chose a professional golfer for the second. As a bit of a stretch, I chose a compiler writer for the third because he was a friend of my father's. That choice from computer science would be prophetic—years later, I ended up writing some compilers, too.

I found the college application process to be very stressful. Since I wasn't sure what direction I would take, I looked for schools with a variety of paths. I was drawn to the community provided by the residential college system at Yale, so I applied there as an early-action candidate. After I was accepted, the only other school I applied to was Stanford, which my parents had attended.

When I visited both schools, I preferred Stanford. Perhaps it was because I was from the West Coast and fit into the culture, but the school felt more friendly and collegial. I knew that Stanford would be a good place to study computer science or writing or any other topic I took an interest in.

A COMPUTER SCIENTIST

The social supports and culture of Stanford were vital for me. Friends at other universities gave me the impression that if they weren't working hard all the time, they were considered to be slacking off. But at Stanford hanging out with friends, relaxing, and doing outdoor activities seemed to be just as important as academics.

There was one aspect of Stanford's social environment that made me feel out of place. Most students were either "techies," focused on science, technology, engineering, or math fields, or "fuzzies," focused on the humanities. With a strong interest in both areas, I didn't fit comfortably into either group. I found technical problem sets easy but not rewarding, and the content of writing essays rewarding but not fun to create. The classes I enjoyed the most were new to me and not taught in high school: geology, religious history, psychology—and computer science.

In the spring of my freshman year, I took a computer science class and loved it. Not only did I fly through the problem sets, but I also thrilled at how I could make the computer do what I wanted. Writing a program was fun, like playing a game. After six hours, I would still be sitting at the computer, actively engaged.

It was in my second computer science class the next fall that I met Steve, who introduced me to the halting problem. One of the ways a computer program can fail is to go into an infinite loop and stop listening to input. That's obviously not what you want, so a tool that can tell you if a program is guaranteed to "halt" or whether there is a possibility it might run forever would be incredibly useful. But it turns out there is no such tool—not because no one has discovered it, not because you would need bigger or faster computers, but because it's an impossibility. Once Steve showed me this proof, I knew I wanted to be a computer scientist. Soon after that, I also knew I wanted to become a computer science professor.

At Stanford, some students who complete the introductory programming class have the opportunity to return as section leaders and help teach
the course. I applied to be a section leader, was accepted, and took training
on how to teach sections consisting of between 10 and 20 students. Later
I became a coordinator for the program, hiring and training other section
leaders. Section leaders attend weekly meetings with the lecturers to discuss
what the lecturers would be teaching that week and what they wanted leaders to cover in our sections. By junior year, I was living with Steve and his
housemates, all of whom were computer science instructors.

In addition to teaching sections, I took advantage of internship and
research opportunities during college. I worked one summer at Apple, where
I helped to write a word-processing program for internal use. I did research
with Professor Jeffrey Ullman on a logic-based programming language called
NAIL! and designed for programs that make heavy use of information stored
in databases.

Between my work as a section leader, conversations with Steve and his
housemates, and experiences working with Professor Ullman, I had a pretty
good idea what being in academia was like, and I decided that was where I
wanted to be.

Choosing where to go to graduate school was complicated. Steve and I
were already married, and he was working nearby at Apple. So financially
it made sense to continue studying at Stanford. But applying to only one
top-ranked graduate school is risky. Stanford has a program that allows you
to spend a fifth year in a department and get a master's degree. I applied to
the PhD program with a fallback plan: if I were not accepted, I would enter
the master's program and then apply to other schools the following year.
I was accepted, though, and continued as a PhD student in the Stanford
Computer Science Department.

A DIFFERENT WORLD

Even though I didn't change schools after college, Stanford felt like a totally
different world to a PhD student. My life as an undergrad was structured
around friends and course work. But now that I was in grad school, my life
was centered in the Computer Science Department doing research. My first
year there was a temptation to get wrapped up in taking classes and not
buckle down on a research project. I was still working like an undergraduate.

I made the transition the second year, thanks to an unusual type of
encouragement—a child. My eventual adviser, John Mitchell, started recruiting me soon after I told him I was pregnant, which I took as a good sign. After

Kathleen Fisher received her PhD in 1996 from Stanford University for her
work on type systems for object-oriented languages.

Elaine was born, John remained extremely supportive—he respected my time
constraints, and his wife lent me some of her old maternity clothes.

Doing research while I had a young child at home made for a very different grad school experience than most of my cohort had. Thanks to an au
pair, I had 45 hours of childcare each week, during which I had to fit in all
my classes, research, and teaching. This meant I got into work at 9:00 each
morning and had three hours of productive time to myself before the rest
of the group showed up. These colleagues, without a child to focus them
and constrain their time, found it much easier just to hang out at work—
playing ping-pong, getting distracted by classes, and staying up until mid-
night. But I went home to dinner every day.

One of the hardest parts of graduate school is making the transition
from defining yourself through regular feedback from classes, grades, and
peers to defining yourself through difficult, often intractable research questions. We were working on problems for which none of us knew the answer.
Without the regular input of good grades and accomplishments, some of
my colleagues begin grappling with existential angst, questioning the value
of their research and themselves.

But because of my daughter, I had more eggs in my basket. Taking care of
her could often be a source of stress, so coming in to work was a relief. And

better, my uncertain research wasn't how I defined myself. For me, graduate school was never an existential question—it was a job. Deciding whether to have kids during graduate school is an extremely personal decision, and no time is perfect, but doing so worked for me. It helped that my husband at the time, Steve, and my adviser, John, were always supportive of my career and my family.

DEFERRING THE ACADEMIC DREAM

Under John Mitchell, I was studying the underlying theory of type systems. These formal systems distinguish between programs that are well formed and those that are guaranteed to fail. One of the ways programs can fail is that they create a value of a certain kind—an integer, a string, a data type that represents a person—and then a later part of the program treats it as a different kind of data. When this happens, you lose guarantees about the program's function and create vulnerabilities. My thesis proved that a particular type system for a C++-style programming language was "type safe": any program that passed the type checker would execute without certain runtime errors.

I hoped to become a computer science professor, so I spent time gaining experience in the nonresearch details of academia: I taught John's 120-student computer science course when he was in England on sabbatical, and I served on faculty hiring and administrative committees. In my last year of grad school, I sat down with my adviser to work out a list of the best computer science jobs in academia to apply for.

At that point, however, strange though it may seem given today's job market, there wasn't a high demand for computer science professors. This was especially true for those with my rather abstract and theoretical expertise. And there was another constraint as well—both Steve and I were applying for jobs, so we had to find two jobs in the same location. When a friend at AT&T called and asked me to apply there (using essentially the same application as for the academic jobs), both Steve and I did so.

I made quite a splash at AT&T that fall by giving a talk in which I claimed that the creators of C++ had overlooked a fundamental flaw in the language's structure. At that point, someone in the back stood up and said, "You're lying!" I held my ground, changing my statement to say that the designers had made a mistake rather than overlooked the flaw. I was informed afterward that the interruption had come from Bjarne Stroustrup, one of the creators of C++.

The next time Bjarne spoke to me, it was to offer me a job in a programming language research group being formed at the newly created AT&T

Labs. It was a strange time to join the labs because AT&T was in the process of splitting into parts, one of which retained the name AT&T and one of which became Lucent. The famous Bell Labs was also split. The part that went to AT&T became AT&T Labs, while the part that stayed at Lucent retained the name Bell Labs. Between my offer and my arrival in New Jersey, the group I was to join decided to remain part of Bell Labs. So I started at AT&T with nobody in the organization particularly wanting me there—a difficult work situation, to say the least.

The first years at AT&T were tough. I moved my research area from theoretical programming languages to domain-specific languages relevant to AT&T. One language I helped create was dedicated to managing massive amounts of streaming data, so the company's statisticians could detect phone fraud. Another was fine-tuned to digest the ad hoc data files that the company's many systems were constantly producing in order to extract useful information quickly.

The work was interesting, and I had a knack for it, but I seemed always to be behind on what my managers expected from me. My colleague, Anne, and I were very careful to give each other full credit, but inside AT&T there was this peculiar perception that she was responsible for all the creative thinking, while I was just carrying out her ideas, which harmed my evaluations even after I found my footing. Strangely, outside AT&T the perception was exactly reversed!

To AT&T's credit, when Anne took another job and I started working on the ad hoc file project without her, the managers suddenly changed their perspective. "Oh, all these years when Anne and Kathleen were saying they were partners, they were right!" That was frustrating at first but amusing by the end of my time there.

My husband, Steve, never fully adjusted to life in New Jersey. We divorced and then both of us remarried, bringing the size of our little family unit up to five people with three incomes—all tied to AT&T labs. By the late 1990s, AT&T wasn't doing very well, so when a headhunter at Google reached out to me, I figured it might be a good chance to diversify the family incomes. I told Google I would interview with them—but only if they also considered my husband, Bob, and my ex-husband, Steve. It was an unusual request from a family with unusual constraints and commitments.

Bob ended up taking a job at Google, while Steve went to Salesforce. When I got my offer from Google, I told AT&T I was moving to California, and they could either let me work remotely or lose me to Google as well. Both AT&T and I preferred that I work remotely, so I worked from California for six years, until Elaine graduated from high school.

RETURNING TO ACADEMIA

Once my daughter went to college, my husband and I were free to relocate. I knew that if I wanted to become a professor, it was the right time—if I were to wait too long, a university wouldn't be interested in hiring me because I would be very close to retirement.

I applied to schools close to places where Google had engineering offices and thus to where Bob could work. Tufts University, located in the Boston area, made me an offer of a full professorship. Tufts has an expedited tenure process for professors hired from other tenured positions, but for me, coming from industry, the tenure review took six months. During that process, I stayed at AT&T in case the position at Tufts fell through.

During this limbo period, I was approached by Peter Lee, an office director at the Defense Advanced Research Projects Agency (DARPA) who had been charged by the agency's new director, Regina Dugan, to improve the agency's connections with the academic computer science community. Peter was looking for program managers who would be able to "speak academia" and thus collaborate with both academic and industrial teams. The

After Kathleen Fisher's daughter, Elaine, graduated from high school in 2010, Kathleen decided it was the right time to move from industry to academia and become a professor. *From left*, Kathleen's husband, Bob Gruber; Kathleen; Elaine; Elaine's stepmother, Sue Fisher; and Elaine's father, Steve Fisher.

Kathleen Fisher is currently chair of the Computer Science Department at Tufts University.

position would be a three-year appointment, during which time DARPA would pay Tufts for my services.

I was concerned that Tufts would not want me to take three years off before I even started working as a professor, but when I asked the dean, she liked the idea. She said my job at DARPA would increase the school's research profile. She also said that an established professor would have trouble taking such a position because it required stalling the PhD student/grant application pipeline for several years.

I found DARPA to be an amazing organization, very effectively and efficiently focused on achieving its mission. I worked on two projects there. The first focused on using what are called formal methods to build software—High-Assurance Cyber Military Systems (HACMS)—for vehicles that is much harder to hack into. At the beginning of the program, our "red team" was able to hack into a quadcopter drone and fly it off into the sunset, but after 16 months of work by the HACMS "blue team," the red team couldn't break in, thus making the resulting quadcopter "the most secure unmanned aerial vehicle on the planet," according to one of my colleagues at DARPA. The program was very successful and great fun as well.

The second project I led involved writing programming languages to make it easier to build machine-learning applications, which was also successful and great fun.

After three years at DARPA, I returned to Tufts, finally, to become a professor. Since 2013, I have been teaching undergraduates (in addition to all my other responsibilities), hoping to inspire them, just as I was inspired all those years ago.

ABOUT THE SUBJECT

Kathleen Fisher is professor in and chair of the Computer Science Department at Tufts University. She was previously a principal member of the technical staff at AT&T Labs Research and a consulting faculty member in the Computer Science Department at Stanford University. She has an undergraduate degree in mathematics and computation and a PhD in computer science from Stanford University, where she was a Hertz Fellow.

19 A FLASH OF INSPIRATION

Tamara Doering

As a postdoc in Randy Schekman's lab at the University of California, Berkeley, I was having a hard time—my experiments weren't working. I was doing basic research in a model yeast system. In graduate school, I had studied the biosynthesis of glycolipid structures that anchor proteins in membranes, and for my postdoc I was investigating how this modification affects protein traffic inside the cell. My graduate work had been very successful, and now I felt stuck by comparison—under time pressure with no clear experimental path to completion. Meanwhile, my future husband kept asking when I would be able to move back to the East Coast to join him. I told him that to apply for an academic position I needed more experimental results. So I kept at it, working long hours in the lab. I eventually got results that I could publish in a strong journal and decided I had accomplished enough to apply for faculty jobs.

To apply for a faculty position, I needed to propose a research area and experimental plan. Applicants usually propose working on something related to their postdoctoral studies to get off to a running start in their own labs. The problem was that I didn't want to continue to do basic research in model yeast, which is what I had been doing. I wanted instead to study an organism that directly causes disease. Pathogenic microbes offer both fascinating basic research questions and direct connection to human health, a compelling combination. But I had no idea which organism to study.

While chatting with colleagues at a conference, I suddenly had a flash of inspiration. I could go back to the topic of my graduate work, the glycolipid

Tamara Doering attended college and graduate school at Johns Hopkins University. She gave a ride to Matt and Rachel Shapiro, children of her graduate school mentor Terry Shapiro, in 1991.

anchors, but study them in yeast, specifically in disease-causing yeast, which I remembered having heard about in medical school. So this almost completely formed new career direction came into my mind: I was going to study pathogenic fungi.

Returning from the conference, I headed to the Berkeley library, read reviews about pathogenic yeast, and chose the organism that I still study. More than 20 years later I am still working on an organism I picked out of a book in the library for all the wrong reasons. At the time, I thought that one could manipulate these yeast in a fashion similar to the model yeast that I had been working on. I thought, "That's easy. I can do that." But that turned out not to be true at the time—the genetic tools and strains did not yet exist. I'm happy to say, though, that I eventually found the right reasons to study this organism, and that work has kept me busy ever since.

AN ACADEMIC FAMILY

I grew up in an academic family in Baltimore, Maryland. My father taught chemistry at Johns Hopkins University, and my mother was a sociologist in academia and government. I have an older sister and a younger brother, and our dinner conversations often related to science. I grew up in a bit of a bubble, where I always assumed that I would go on to get an advanced degree after college and become a research scientist; in our family, this plan was assumed, and my siblings and I followed career paths in science and medicine.

I attended an excellent private school for middle and high school. It was a progressive school with open classrooms and individualized learning. The teachers were great, and two of my favorite subjects were literature and history. Although the school was oriented toward the liberal arts, I also took science courses, including biology, physics, and chemistry, as well as math through calculus. I remember the biology classes as being particularly well taught. And although I really enjoyed the liberal arts subjects, I never really considered a career in that direction. As I was growing up, my role models were scientists, so that's what I decided to be, too.

One choice Michael and I made was to delay having children until later in our careers. A lot of students ask me about the best time to start having a family. I tell them that having kids is amazing, and there is no right time, just a mix of advantages and disadvantages. Some scientists have children during their graduate school and postdoc years. If I had done so, I wouldn't necessarily be where I am now professionally. There were a few key years when I was getting established in the field. If I had been struggling to launch my research together with the extra job of being a mother, I don't know that I could have succeeded.

An advantage of having kids later in our careers was having financial resources that gave us flexibility—for example, for additional childcare. Terry Shapiro, a wonderful mentor I had at Hopkins, once told me that my ability to succeed in my career would be directly related to the quality of the childcare I could find. She was absolutely right—if I'm worrying about my kids or running out to take care of them, I can't focus on my professional obligations.

A disadvantage of waiting to have kids is that we're older parents. Younger parents can keep up with their kids longer and share more of their lives with them. Also, our own parents are older and can't be engaged as actively with their grandchildren. But it all works out either way—our kids keep us young and active, too. In the end, family is more important than career. If someone feels like it's the right time in her or his life to have children, I would never suggest otherwise. I think you can always make it work.

I wouldn't change the way my career and family path played out. I enjoyed being a postdoc, living in the San Francisco Bay Area with no family commitments. It was a wonderful time of my life; I got to have fun in a wonderful part of the country and make friendships that still sustain me. And carving out the large number of hours I worked in a lab as a postdoc would have been much more challenging with a family, although certainly people do manage this successfully.

I turned 16 a few months before I graduated from high school and didn't feel ready to leave home. With my dad teaching at Johns Hopkins, I considered it the local school and didn't apply anywhere else for college. I commuted from home with my older sister for my first two college years and with my younger brother for my last two years. Going to college with my siblings was really fun. My older sister was a biophysics major, and I also started in that department. I loved an evolutionary biology class taught by Professor Michael Edidin, and at the end of the semester I switched my major to biology and asked him for a summer job. At the time, he had a couple of senior students in his group who were finishing up their PhD work. I helped them with their projects, and they taught me about how to work in a lab. Starting that summer and until now, I have always been associated with a biology lab.

A CAREER IMPRINT

After sophomore year, I asked Edidin for another summer job, but he was leaving for a sabbatical. So I walked down the hall to the next lab and asked the professor there for a job. That professor turned out to be Saul Roseman, a pioneer in glycobiology, the study of carbohydrates. Roseman offered me a job, and I had a great experience in his lab. I continued to work there for three years during college and for another year after I graduated. I feel as if the direction of my career was "imprinted" by the work I did for him. Glycobiology has been a unifying theme in my research since that summer job my sophomore year in college—although this hasn't been intentional.

The Roseman lab was a large place with an international group of researchers. During my four years there, I worked with several graduate students and postdoc, but two stand out as mentors. A postdoc from Scotland, Wilf Mitchell, taught me to do meticulous biochemical assays and about the game of rugby. I also worked with David Saffen, the first graduate student to bring molecular biology approaches into the Roseman lab, which had always been known for biochemistry. David and I worked together to sequence DNA at a time when the process was not yet automated. We sequenced just a few hundred base pairs, in contrast to the millions that are done routinely today.

I enjoyed interacting with Professor Roseman, who was a brilliant biochemist and an exacting supervisor. I grew up in a faculty household, so I wasn't afraid of an autocratic professor, which helped. Roseman liked my outspoken style and my hard work, and he really encouraged me and advanced my career. Also, as an undergraduate student, I had a privileged position—I got to learn biology and have fun doing experiments without the demands and stress of completing a degree.

From an early age, I had assumed that I would go to graduate school after college. When I was about to apply, my dad asked if it was really what I wanted to do. I became upset and defensive and said of course it was what I wanted. But after that conversation I realized I had never thought carefully about my career direction, so I decided to wait another year before applying.

GRADUATE SCHOOL OR MEDICAL SCHOOL?

Many of my friends at Johns Hopkins were premed students, and I had heard a lot about their plans for medical school. As a freshman, I had considered the idea of entering an MD/PhD program after college but was discouraged by a faculty member who felt my grades wouldn't be good enough, so I had never followed up on the idea. My senior year I started thinking about the idea again and went to see Paul Talalay, who was the head of the Hopkins MD/PhD program. He was very encouraging, and I decided it was the right career direction for me. I spent the year after graduation working in the Roseman lab and taking a few remaining prerequisite courses. I applied to several programs but in the end decided to stay at Hopkins.

After my first year of medical courses, I needed to pick a graduate lab rotation for the summer. I had read about the work of Paul Englund, who was studying the parasite that causes African sleeping sickness, and the research sounded fascinating. Paul told me that he was going to be running a 10-week summer course at Woods Hole on the biology of parasites and invited me to apply. The course was a wonderful, intensive introduction to parasitology that combined lab and classwork and introduced me to terrific scientists in the field. Based on that summer, I decided to do research in Paul's lab.

Once I completed the first two years of medical school, which included courses and several hospital rotations, I began work in both Paul Englund's lab and a lab headed by Gerald Hart. The two groups were collaborating to determine how a newly discovered molecule that anchors proteins to biological membranes was synthesized. The anchor molecule is very abundant in the trypanosome parasites studied in Paul's lab and is composed primarily of carbohydrates, the specialty of the Hart lab. We figured out how the parasite constructs this unusual structure, which is partly lipids and partly sugars. The project was exciting and extremely successful research in a new high-profile area, so our work received a great deal of attention. I got to present our research at conferences and even chaired a conference session, a level of recognition not typically given a graduate student.

Paul was a wonderful mentor and remains my strongest scientific influence. He taught me clear thinking and communication—how to strip a

Tamara Doering was a postdoc in Professor Randy Schekman's lab at the University of California—Berkeley in 1996.

presentation to its essentials and how to present data. I still use his approaches when I write a paper or reply to a critique from a referee. Paul was also so excited about science that it was contagious. At the time I was working with Paul, molecular biologists collecting data developed their own X-ray film using photographic chemicals in a dark room. A guaranteed way for a student to get Paul's attention was to rush out of the dark room with a dripping-wet film. For fresh data, Paul would drop absolutely everything, forgetting about schedules and meetings, in order to discuss the new results. I also learned a tremendous amount from my other thesis mentor, Jerry Hart. He shared with Paul an absolute love of science, a knack for asking the right questions, and the ability to come up with creative ways to answer them.

MEDICAL PRACTICE OR A RESEARCH CAREER?

After completing my thesis work, I had to finish my medical school training. Although I loved interacting with patients, I realized I wanted to stay in basic research for my career. My choice was related in part to changes in health care that had occurred during my years in the lab. By the time I went back to medical rotations in 1990, many of the patients in the Baltimore

Tamara Doering (*right*) and her research group moved into new laboratory space at the University of Washington Medical School in 2001.

hospital where I worked were HIV positive, which at that time was pretty much a death sentence. I felt that this situation changed the atmosphere in the hospital and the way medical staff interacted with patients. Health-care costs had also become a much larger issue, which further changed clinical decision making. Finally, I was bothered by the strict hierarchy and gender bias I encountered at the Johns Hopkins Hospital, which was a conservative, male-dominated place. As I noticed all of this, I realized I was also missing the excitement of science and discovery in the lab. So even though I loved seeing the patients and found medicine fascinating, it was not the environment I wanted to be in long term. Instead, after finishing the required medical school rotations, I started looking at possible mentors for postdoctoral work. I asked scientists on the campus and read biomedical journals for ideas, visited labs in various fields, and eventually chose to go to Randy Schekman's lab at Berkeley.

The Schekman lab was studying yeast intracellular traffic, which is how proteins, once synthesized, make their way between intracellular compartments and out of cells. After an initial project that didn't pan out, I developed a project related to the work on glycolipid anchors that I had done in graduate school, looking at how these anchors affect protein movement in the yeast cell. The lab had amazing people and resources, but my project was not a mainstream research question for the group, and I didn't have much background in yeast intracellular traffic. I was working hard, but my experiments weren't progressing, and I got a bit discouraged. At one point, I talked to a company that was interested in recruiting me, and I went to a few talks about other career paths for PhDs, such as management consulting. I never actively pursued those directions, but I was definitely concerned about how I could be successful in science.

Meanwhile, my future husband, Michael, was on the East Coast and wondering when I would return there and join him. We had met as teenagers and lost touch but then got together again when he moved to Baltimore to take a faculty position at Johns Hopkins. He was worried because he had expected my postdoc position in California to last a couple of years, but after two years I hadn't even started applying for positions, and I had told him that I couldn't get a job unless I could get my project to move forward. At that time, there were 14 postdocs in the Schekman lab, and I was watching the ones who were a year or two ahead of me and looking for jobs. They were finding positions, but I could see the process was challenging and stressful. Until then, I hadn't really thought about how to get a faculty position, especially how to get one where Michael and I could be together, or about what I would work on. And, of course, to do any of this, I needed to propose an area of research.

> Even with really well-meaning partners, work–family balance issues fall harder on women. There are still double standards in academia about men and women. Some of it is unconscious bias on the part of both genders. If a woman leaves a meeting to pick up her kid, people think she's not attending to her work. But if a man leaves a meeting for the same reason, people think what a great dad he is. Women are still working toward professional and pay equity. So for women in science with families, there is a lot of juggling and balancing, but I wouldn't give up either role.

I distinctly remember the day when I figured out what I would work on for the rest of my career (at least so far). I was in the back of the room at a conference, chatting with a colleague, when I got the inspiration that I should work on yeast that cause diseases. This would be the perfect combination of my experience with pathogenic microbes in graduate school and my new skills in studying yeasts. I was really under pressure and feeling rather desperate about a research direction that would inspire me when this idea came along and saved me. Returning home from the conference, I went to the Berkeley library and read reviews on pathogenic yeast. To do the type of genetic studies that I had learned about in the Schekman lab, it's a big advantage to have an organism that can be induced to undergo genetic recombination to make progeny rather than just dividing and making exactly the same organism again. I picked *Cryptococcus* because I thought these manipulations would be straightforward, so I could apply the power of model yeast genetics to the study of the glycolipid anchors of proteins in this organism. It turned out, though, that neither of these ideas was practical, at least at that time, but I eventually moved on to interesting questions about the organism that I have worked on ever since. For example, *Cryptococcus* has a big polysaccharide capsule that is required for it to cause disease. How does it make that capsule? How does it regulate the capsule's size and nature in response to particular environmental conditions, such as life in a mammal? How does the fungal cell interact with host cells, and how does it get into the brain, where it causes fatal disease? There is all this amazing biology, none of which can be studied using model (rather than pathogenic) yeast. Those are the sorts of questions my group has been studying for the past 20 years.

That one idea, to study glycolipid anchors in pathogenic yeast, pulled together all the strands from every training experience I had had, from my undergraduate work on. In a way, it was a very logical synthesis of those experiences, although at the time it felt like an idea that came out of nowhere. Today it would be hard to convince a sponsor to fund a young researcher with such an idea. I was a postdoc saying that I was going to change fields, start a lab, and work on an organism that I had never touched before. It's remarkable that Cornell Medical School believed that story enough to hire me to do it based primarily on the strength of my record from graduate school. And the Burroughs Welcome Fund believed in me enough to give me a career award. Those two things allowed me to continue down the path to become an academic research scientist.

Michael and I got engaged in late 1996, just before I moved back east fresh from completing my postdoc. I deferred a faculty position at Cornell Medical School until the following summer and spent the interim as a visiting scientist in a lab that worked on *Cryptococcus*. Michael took a sabbatical from Hopkins so we could live together in New York. He and I then undertook a big job search so we could be together and have good jobs in the same city. That search was very stressful since there were no guarantees that we would find something good for each of us. We were fortunate, though, and ended up with a choice of positions at two or three great universities. We thought hard about what was the best choice for us as a couple—if each of us had been on our own, we might have chosen differently. The following summer we moved to St. Louis to start our labs at Washington University and have been there since. It was a terrific choice—the city, the science, and the people have exceeded our hopes, and we have been very happy here.

ABOUT THE SUBJECT

Tamara Doering is the Alumni Endowed Professor of Molecular Microbiology at Washington University Medical School in St. Louis, Missouri. She has a BA in biology from John Hopkins University and an MD/PhD from the Johns Hopkins School of Medicine.

20 A CHANGE OF DIRECTION

Michael O'Hanlon

It was a moment every teacher lives for. Ten high school students knocked on my door. "Will you please teach us? We really need a professor, and no one else will!"

Having had a predicable life thus far, I had decided to join the Peace Corps. I found myself in Zaire, the heart of Africa, thinking that I could make a difference in the lives of disadvantaged students there. They desperately wanted to learn, to get some key to a better life. And yet, despite their eagerness to learn and my eagerness to teach, at some point on my daily commute across a river of hippopotamuses, I realized that there was nothing I could

do to help them. Zaire was a pawn in a Cold War struggle, dominated by a dictator (Mobutu Sese Seko). There was no infrastructure, there were no opportunities. The global powers were worried about geostrategic alignment, not about good governance. I blame the dictator most for this situation, but I blame us, too.

As I sat in my hut in the Central African rain forest, it seemed appealing to go back to a place I knew well. I had studied physics as an undergraduate at Princeton, and Princeton was happy to have me back. I enrolled as a PhD student in the Environmental and Energy Program of the Mechanical and Aerospace Engineering Department. Some years later I emerged with a PhD in policy and a career as a policy analyst.

We write our résumés listing our careers as if they were preordained, with excellence and clarity of purpose. Sometimes I feel like writing an alternate résumé that would list all the ways I struggled and still do. It would high-light the two times that I almost flunked out of grad school. It would include my senior thesis with famous physicist Robert Dicke. It was great working with Dicke, studying solar oscillations and trying to disprove Ein-stein's theory of relativity. I imagine putting on my alternate résumé, "Turns out that Einstein was right and we were wrong."

LEARNING AND TEACHING PHYSICS

Both my actual and alternate résumés begin in Canandaigua, New York, the small farming town in the Finger Lakes region of rural upstate New York where I grew up. My grandfather worked at RCA on early radio and radar systems for the space program. He passed away in 1977, when I was 16, leav-ing me with the inspiration he provided as the main person in my family inclined toward physical sciences.

Dad had an intellectual streak but was also self-deprecating. He was a physician, but he never would have described the sort of medicine that he practiced as overly sophisticated or scientific. Later, when I was starting to do well in math and science and tossed out the idea of going to medi-cal school, Dad would say, "Ah, don't waste your brain on medicine." He lifted my sights with his respect for pure scholarship and made it seem socially and politically acceptable, even meritorious. That said, I think he was wrong about medicine, and I am incredibly proud of both him and my older sister, who followed in his footsteps.

I spent my summers on the farm of one of my science teachers, working with livestock and wildlife. He was a born naturalist and ran the school's ecology club. Ironically, it was through my work with soil and machinery for that club that I was first exposed to physics. In the summer of 1976, I attended a sophisticated program on astronomy at Phillips Academy Ando-ver and was immediately hooked.

For two years I attended Hamilton College, a liberal arts school in Upstate New York not too far from my hometown. Hamilton was great, but, truth be told, no other school that I had applied to would accept me. I had decided to graduate from high school in three years, so although I had good SAT scores and some advanced-placement courses, I didn't have the requisite set of completed courses that many top schools required. I applied to one Ivy League school, Dartmouth, and was turned down. I applied for a postgraduate year at Andover Academy, where I had taken the summer program, but it, too, turned me down. Some of my friends from Canandaigua had gone to Hamilton, and Hamilton accepted me, so that's where I enrolled.

There were wonderful physics professors at Hamilton. Although the faculty was small, the professors there—in particular Jim Ring and Phil Pearl—taught me about the beauty and simplicity of physics, astronomy, astrophysics, and the creation of the universe. By the end of my second year, though, I had exhausted Hamilton's physics offerings and couldn't find peers to match my level of passion for this science. So I transferred to Princeton, lucky to have applied in a year when it was still taking transfer students. But the teaching never got better, not even at Princeton.

I feel as if I took college physics at Hamilton and graduate physics at Princeton. That's no slight on Hamilton, just a reflection of my own level of understanding when I was at each place. I left Hamilton with a range of physics courses under my belt, but when I arrived at Princeton, I was told to do my sophomore year over again. Undeterred, I took a difficult quantum mechanics course from Will Happer and did pretty well, and the school decided that I need not to repeat a year after all. I continued to succeed, more or less, throughout my time at Princeton, even through a full year of graduate quantum mechanics with string theorist Ed Witten. I also enjoyed experimental work. The experimental physics lab course taught by biophysicist Bob Austin was one of my most memorable experiences.

I slightly overdid it in college, with lots of physics at Hamilton, then lots physics and math at Princeton. By the end of senior year, I was tired of physics but still not yet ready for graduate school. My interests in history and global affairs, along with an emotional and moral imperative to do something for those who were not as fortunate, led me to join the Peace Corps, to teach physics at the college level in Kikwit, Zaire. I was lucky in that the Princeton language requirement had forced me to learn French at a high-enough level that I could really communicate. In addition to teaching, I became involved in a local project, working with a Catholic priest to improve the cleanliness of drinking water taken from a local subterranean spring.

I am still glad I went into the Peace Corps and tried to help, but at the time there was not much any of us could do to help Zaire, at least not in broad terms. I decided to return to the United States to continue my studies in a program that combined science and public affairs. I chose the Energy and Environment Program in the Princeton Mechanical and Aerospace Engineering Department.

After Michael O'Hanlon graduated from Princeton University in 1982, he decided to join the Peace Corps. He spent a year teaching in Zaire, where the students were eager to learn physics.

Michael O'Hanlon's daily commute to class in Zaire included fording a river, sometimes past bathing hippopotamuses.

From his hut in the Central African rain forest, it seemed appealing to Michael O'Hanlon to go back to the United States for graduate school at a place he knew well. He returned to Princeton, where he completed a master's degree in engineering before switching fields and earning a PhD in public and international affairs.

SWITCHING FIELDS IN GRADUATE SCHOOL

Once back at Princeton, I struggled. This was the Reagan era, and I worked with Frank von Hippel to model Soviet air defenses and ballistic missiles, but my departmental adviser and the faculty on my committee didn't think the topic was sophisticated enough for a PhD. They were right. It wasn't an experimental thesis—we certainly didn't have any Soviet air-defense equipment to play with! And the physics was at the undergraduate level. The faculty told me to throw away that thesis topic and start over. That was the first time I almost flunked out of graduate school.

I didn't want to waste a year of work, so the modeling became a master's thesis. Like me, von Hippel had an interest in public affairs and had an appointment at Princeton's Woodrow Wilson School of Public and International Affairs. With his support, I transferred to a PhD program at Woodrow Wilson. The difference I observed between the social scientists and the physical scientists was stark. Whereas the physicist tries to simplify as much as possible, the social scientist seems to want to complicate things just for the sake of complication. But there were lots of great courses, too.

Policy analysis is difficult because you have to combine different kinds of information from different areas just to get analytical traction in a given problem. It can be about Machiavelli and what he wrote half a millennium ago or about technology and current weapons systems or about what drives China's leaders in the twenty-first century. The work is important and just as difficult as pure science. But as a graduate student with a skeptical attitude toward the field, I ended up flunking the political science portion of the general exam. I needed to be "rehabilitated." I struggled but managed to pass the exam with help from Professors Dick Ullman and Hal Feiveson, among others. That was the second time I almost flunked out of graduate school.

My dissertation, supervised by Josh Epstein and Aaron Friedberg, blended a little bit of science, a lot of policy analysis, and military analysis. It was about new ways to think about defense policy as a nation at a time the Cold War was winding down, a topic I later continued examining as an analyst at the Congressional Budget Office (CBO) and indeed ever since.

A MILITARY POLICY ANALYST

After completing my PhD, I was a summer intern at the Institute for Defense Analyses, a military think tank. From there, I went to the CBO in the fall of 1989, where I spent five fruitful years thinking about new ideas in military defense policy. That was a wonderful period for new ideas. The Cold War had just ended. I was technically a nuclear weapons analyst and traveled to the Department of Energy Laboratories to understand how we could maintain the reliability of nuclear weapons as we entered an era without nuclear testing. We worked on rescaling to a smaller nuclear weapons complex for the post–Cold War world.

I also did some studies relating to Middle Eastern security. Operation Desert Storm happened during my period at the CBO, but before the United States began the operation, we were asked how much the war might cost. Our estimate was that it would cost between $50 billion and $150 billion. The actual figure turned out to be $100 billion. In reality, though, we never tried to be too precise in our cost estimates for a conflict. When someone would ask us to be more precise, we would say that we don't have crystal balls, so a precise estimate would likely be wrong. We would also say that when an estimate turns out to be correct, it will likely be by luck. In the history of warfare, the ability to forecast the nature of a conflict before it happens is extraordinarily unusual and is always unwise.

In 1994, I left the CBO to join the Brookings Institution, and I have been there ever since. At Brookings, we applied to the Iraq War of 2003 the same sensibilities used at the CBO. I forecast the nature of the conflict, and I had about a factor of 10 disparity between my lower casualty estimate and my higher casualty estimate for an invasion of Iraq. I turned out to be more or less correct. The casualties for the United States slightly exceeded my upper bound, but at least I conveyed the range of the plausible.

Brookings was always my dream job. I had read Brookings literature when I was in graduate school. I thought it was the place generating the most policy-relevant and serious research of any place in the country on the topics that I cared about. There were good people at MIT and Johns Hopkins and at other think tanks as well. But in the 1980s Brookings was exceptional, with six or eight spectacular scholars, including my PhD adviser Josh Epstein, who commuted from Washington, DC, to teach courses at Princeton as an adjunct professor, as I do now. Josh Epstein has had an enormous influence on me—on both my graduate school and my Washington experiences. He helped with the job applications for the CBO and Brookings. He taught me how to write crisply. He is one of the most important role models I have ever had.

After Josh Epstein, another crucial influence was Professor Aaron Friedberg. He was the perfect dissertation adviser. Unlike some professors, he never wanted to turn my thesis into a seven-year adventure. His goal was to help crystalize a direction early in the project. Friedberg met me a few times to make sure I was on track and gave me a few thoughts along the way. He read my entire draft thesis in one short period of time; then he wrote a five-page memo about what should be in the second draft to improve it. I followed his advice, and we were done! I like to think that the quality was good, too, but the process was wonderful.

Aaron should give a course on how to be a good thesis adviser because not many professors understand the process as well as he does. One's time in graduate school is not to be frittered away just because an adviser decides he or she has the power keep you waiting. A poor adviser may see graduate school as a rite of passage, but I find that attitude abusive. The adviser's loyalty to the student should come first. If the student is working hard and making the headway that the field expects, the adviser should help the student finish quickly, have an impact with the study done, and get on with his or her life.

TEXTBOOKS AND OP-ED PIECES

While at the Brookings Institution I have written quite a number of books, including textbooks on military policy. I also teach these subjects at Princeton. About a quarter of my military policy books over the years have been on technical subjects—making budgetary calculations, modeling combat, and studying scientific issues in war, such as missile defense. These topics tie together my scientific and policy background from my time at Princeton.

Another quarter of my work concerns mainline US defense strategy and budget. I have written a great deal on regional security issues, in which I usually team up with an expert on the Middle East or East Asia. I try to attack a broad problem so the book will be relevant over a long period of time but also speak to the policy debate of the day. I have written about the wars in Iraq and Afghanistan and about alliances and challenges in East Asia.

In 2014, with Jim Steinberg I wrote *Strategic Reassurance and Resolve*, a book about the US-China relationship. Steinberg is dean of the Maxwell School at Syracuse University and was deputy secretary of state under Hillary Clinton during the first years of the Obama administration. In 2015, I wrote a book called *The Future of Land Warfare*, which is a conceptual book on defense policy. It is meant to be not just about the immediate defense debate but about where conflict is headed and what the United States needs to do to be prepared. I also do a lot of short-term writing, including op-ed pieces about the crises in Syria, Ukraine, and Afghanistan.

SCIENCE AND MILITARY ANALYSIS

Studying natural science shapes the way your brain works for whatever topic you may pursue. It has been very central to what I have done as a scholar in defense and security work. I don't do "Ed Witten–like" quantum physics, but my familiarity with computation and math is essential to what I do. Numerical work with budgets and military modeling come easily to me. With my science background, I never feel pressured to exaggerate the rigor or complexity of my work. I work in terms that are directly accessible to policy makers and achieve greater impact as a result.

Military analysis is largely about technology, and there are three main entries into defense analysis: through military history and broader strategic history; through military service; and, my initial way, through technology. The third way is also the way the nuclear bomb designers got into the arms-control field. Ideally, whatever your means of entry, you want to learn about the other two as well.

Once entering the defense field through one of those ways, you spend the rest of your life reinforcing your strength and gaining knowledge and expertise in the other two. I have spent a lot of time around soldiers and airmen, sailors and marines. I visit them in war zones and at training sites. I have spent time talking to them in Washington, DC. I have also spent a lot of time reading military history. I consider these activities to be ongoing efforts to shore up weaknesses in my résumé. The most important thing for me is to have the curiosity to keep getting better in the areas that are not my strengths.

Since making the decision during grad school to move from science to policy, I have not regretted it. Policy, for better or worse, is what brings us war or peace. The field needs people who can tie many strands of thought together. Many scientists are not good at that, but some are, and they are needed.

ABOUT THE SUBJECT

Michael O'Hanlon is director of research at the Brookings Institution in Washington, DC. He is a scholar and prolific author in the areas of military and strategic policy and international relations. O'Hanlon has an AB degree in physics, a master's degree in mechanical and aerospace engineering, and a PhD in public and international affairs, all from Princeton University.

21 AN UNFAIR ADVANTAGE

David Spergel

There were times when as a student I thought I might pursue a career outside of science. In high school, I was active on the school newspaper and in student government. When I was an undergraduate, my interest in communications and policy continued. During my senior year, I applied both to graduate school and to law school. At graduate school, I wrote a policy paper critically evaluating the *Star Wars* laser-based strategic-defense system. If my PhD work at Harvard had taken longer or been less smooth, my career might have gone in a different direction.

But my graduate work did go smoothly and quickly launched my professional career as a scientist. I finished my dissertation in a little more than two years, and after a two-year postdoc appointment I was hired as a professor by the Princeton Astrophysics Department.

Completing a PhD at Harvard in two years is close to a record. There are a number of reasons why I finished so quickly. I arrived at Harvard after doing research at Princeton and Oxford, so I understood the research process. I also had already completed my graduate course work, so I could concentrate on research. And, as a theorist, I was able to use data generated by others instead of having to build apparatus from scratch. But I had other advantages as well: as an undergraduate at Princeton, I had access to some of the first data from a new large-array radio telescope, and as a graduate student at Harvard I worked in the then new area of dark-matter detection and approached the problem from a different perspective than previous researchers.

David Spergel advised Maggie Thompson for her senior thesis at Princeton University in 2016.

I have found useful an adage from physicist Hans Bethe that I learned from my mentor, John Bahcall, during my postdoc: Bethe said that one should concentrate on the problems for which one has an *unfair advantage*. This means that to make real progress on a well-studied problem, one needs an advantage that others don't have. At both Princeton and Harvard, I had such an advantage.

An unfair advantage can take different forms. It might be a new technique to apply to a problem—possibly a mathematical technique, but more often a piece of technology, such as a novel detector or a new way to make a measurement. Or the advantage might be a theoretical insight from another branch of science—in my case a new idea in particle or condensed-matter physics that I applied to cosmology. I am fortunate. During my career, I have been able to identify and apply unfair advantages to some very interesting scientific problems.

A ROLE MODEL IN THE FAMILY

I grew up in Huntington, New York, on Long Island. I attended the Huntington public schools, which were reasonably good in math and science. There were a number of good teachers in high school, but the teacher to whom I was closest taught journalism. I was an editor and wrote articles for the school newspaper. I learned how to write well in high school—the communication skills I learned have proved very important during my career.

My father was a physics professor at the City University of New York (CUNY) and my mother was a high school home-economics teacher. I grew up in an environment that encouraged me to go into science. Not that my father actively pushed me in that direction—he didn't push me or my siblings very hard or at all. In fact, I was the only one of us to go into physics. But my father provided a role model for what it would be like to be a physics professor. At CUNY, the faculty doesn't do much research; they mostly teach. My father had a really satisfying career as a college teacher, mostly teaching students who were the first in their family to go to college.

At the end of high school, I was accepted into Princeton but not sure what field I would study. I considered math and physics as well as government and policy at the Woodrow Wilson School. I found freshman year at Princeton very challenging. I took the introductory math course for math majors and discovered that although I might have been the strongest math student in my high school, I was not cut out to be a mathematician. I also found that I am uncoordinated in the lab and not suited to be an experimentalist.

Beginning in junior year, I discovered that I really like doing research. I did theoretical work on orbital dynamics with James Binney, a visiting professor from Oxford, during the first semester and worked on radio observations of interstellar media with Gillian Knapp during the second semester. I continued working with Professor Knapp during senior year. For my senior thesis, I studied dying stars using the Very Large Array radio telescope in New Mexico. As stars like our sun die, the outer layers are blown off by strong stellar winds, forming planetary nebulae. We made some of the first observations of these protoplanetary nebulae. The radio telescope was still being built when I visited. I was fortunate enough to obtain some of the first data from the Very Large Array and to publish a paper from that work.

At the end of my senior year, I again was unsure what direction to take. I applied to four graduate schools with astrophysics departments as well as programs in law and government. My top-two choices for graduate schools were Harvard and the California Institute of Technology (Caltech). During a visit to Harvard, I met Professor Bill Press. Bill and I hit it off—we were a good match. We discussed various technical problems, and I decided I would enjoy working with him. So I enrolled at Harvard knowing my likely thesis adviser but realizing that in a large department I would always be able to find good adviser if it turned out not to be Bill Press. I also chose Harvard for graduate school because I expected the social environment in Cambridge to be richer than at Caltech.

After I was accepted to Harvard, I deferred enrollment for a year to go to Oxford. It was a great year—I enjoyed living in a different country and made a lot of friends. While at Oxford, I took a couple of classes but mostly did research and ended the year by writing up a paper on galactic dynamics. After having done research at Princeton and Oxford, I arrived at Harvard in the fall fully prepared and ready to start a thesis.

As I started graduate school, high-energy particle physicists were wondering if they could create dark matter in a high-energy accelerator. Theorists had recently postulated that most of the matter in our universe isn't made of protons or neutrons, which is the type of matter that we are familiar with. They instead claimed dark matter to be weakly interacting massive particles (WIMPs), as predicted by an extension to the standard model of particle physics called "supersymmetry."

As an astrophysicist, I was in the fortunate position to approach the dark-matter question from a different direction. If the particle physicists were correct about the composition of dark matter, I wondered, what other ways could dark matter be detected? At Oxford, I had continued working

David Spergel advised Eiichiniro Komatsu, who completed his PhD thesis work at Princeton University in 2000.

with James Binney, my mentor for the paper I wrote on galactic dynamics in my junior year, so I knew a lot about dynamics and something about the structure of the galaxy.

AN OPEN PLAYING FIELD

In my PhD thesis, I pointed out that because the earth goes around the sun, and the sun moves around the galaxy, the earth's velocity changes with time on an annual basis, moving fastest with respect to the dark matter in the galaxy in June and slowest in December. This annual modulation in velocity means there should be a modulation in the event rate as a function of energy, one that is relatively straightforward to calculate. This modulation has become the observational signature that people have used in their experiments since then to distinguish dark matter from background events. Since I was the first astrophysicist in this area, I had an open playing field with several interesting and relatively easy problems to solve. I quickly wrote a series of papers and combined them into a thesis.

> Being good at research takes a combination of things. First and most important, you need to be interested in research, enjoy it, and want to do it. Second, it takes time and work. A third piece is talent. There are different kinds of talent, different sets of skills that allow scientists to succeed. There are people who are really good at mathematics, and others who are good at computing. Creativity is also very important. Creativity is the ability to find different ways to solve problems.
>
> I consider the most important quality [to be] the ability to identify interesting problems. There is a set of students who are good at solving a problem they are given. The next level is being able to find interesting opportunities. That is the hardest skill to teach as a mentor or a professor, but it is essential for having a big impact.

After finishing the PhD, I remained at Harvard as a postdoc. During graduate school, I received a letter from John Bahcall of the Institute for Advanced Study (IAS) in Princeton inviting me to come for a visit. Bahcall then offered me a position as a long-term IAS member. I was at IAS for two years. It was a good period for me because I moved into new areas. The

Supernova 1987a explosion occurred, and I worked with John to model and understand aftermath details such as neutrino fluxes and cooling. I also worked more on dark matter and its effects on the shape of our galaxy. This work led me to explore the shape of the galaxy. Together with Leo Blitz, a radio astronomer, we showed that the Milky Way is a barred galaxy.

I was happy doing research at IAS, but teaching is important to me, and I knew that I eventually would want to teach. I was collaborating on a project with one of the members of the Princeton Astrophysics Department when department chair Jerry Ostriker encouraged me to apply for a faculty position. I joined the department and have now been a professor for 30 years. Although I'm a theorist, I like to work collaboratively with experimentalists and astronomical observers. I work on smaller projects with one or two others or on larger projects in big groups. I like to work with others and to learn from them. I try to identify novel ways to use some of the planned astrophysical instruments to address interesting questions. And I still look for or create advantages to make progress in new areas.

For example, I have recently become interested in detecting extrasolar planets, or exoplanets, which has required me to learn about optics and optical coronagraphs. Coronagraphs are attachments to telescopes that block out the direct light from a star so that nearby exoplanets can be resolved. I had never learned optics as a student, so I decided to read a standard text and gave myself a homework assignment to design a coronagraph. Using mathematical tricks in Fourier space that I had previously used in cosmology, I was able to visualize a new solution for coronagraph analysis. I am collaborating with colleagues who have experience in optical engineering, and together we have published a number of papers on better ways to detect exoplanets.

Writing and policy are areas that I have always liked and that I considered pursuing as a student. These areas have become more important again in my later career. I am chair of the Space Studies Board of the National Academy of Sciences and on the NASA advisory committee. I serve on many advisory boards and run a couple of big projects. I spend much of my time doing policy and management. The key skills that allow me to do these things are listening to people, writing reasonably well, and stepping back to think about the big picture, separating minutiae from the large-scale problems.

ABOUT THE SUBJECT

David Spergel is an astrophysicist with degrees from Princeton University and Harvard University. He is a professor in the Princeton Astrophysics Department and an associate faculty member in the Physics and Mechanical and Aerospace Engineering Departments. He is also the director of the Center for Computational Astrophysics at the Flatiron Institute in New York City.

22 SOLVING PUZZLES

Shirley Tilghman

I cannot remember a time, going back to my earliest childhood, when I didn't love math. My father tells a story of doing mental math games with me instead of reading me bedtime stories. I'm sure that warped my personality for the rest of my life! He claims it was really my idea, not his. In high school, I became very interested in chemistry—I loved it and went to university thinking I was going to be a chemist.

The university system in Canada is like the British system, where you go in already knowing what you want to do, and you do it very intensively. I was studying chemistry, physics, and math and planning to be a chemist. The only course I took outside of these fields was scientific German, which was not really a broadening experience. But by my third year in 1967, I clearly came to the conclusion that chemistry was not going to be a calling for me. At that point, I started looking around for another scientific discipline to which I might be better suited. I eventually discovered molecular biology.

DISCOVERING MOLECULAR BIOLOGY

Discover is the right word—I grew up in Canada. At the time, if you were a good student, you studied chemistry and physics. If you were a weak student, you studied biology. I was in my third year at the university, and I had never taken a biology course in my entire life. I stumbled upon molecular biology in reading about what has now become one of the most famous experiments—the Meselson-Stahl experiment.[1] Matthew Meselson and Franklin Stahl used isotopes of nitrogen to establish the mechanism by which DNA is replicated. It was the most beautifully designed experiment— just luscious! It addressed and absolutely persuasively answered a really critical question: How is genetic information replicated? It was literally a eureka moment. After reading the paper, I decided that I had better go find out what this DNA stuff is all about.

I signed up for a biochemistry course in my fourth year, thinking I would be able to marry what I knew about chemistry with biology. It was not a particularly good course, and I was lost much of the time in the beginning because I didn't know the language (for example, I didn't know what a phage was). But the course gave me enough of a grasp of molecular biology that I was hooked; this was what I wanted to do.

Coming to the end of my undergraduate studies, I was thinking about going to graduate school, but the only thing I was qualified to study in graduate school was chemistry, which I didn't want to do. So I decided to take two years off and went to Sierra Leone in West Africa with an organization called the Canadian University Services Overseas (CUSO) and taught high school chemistry. CUSO has somewhat the same mission as the Peace Corps, but it is organized differently. My goal was to take a break from academia—I was a little burned out, having worked in labs for three of my four years. I knew if I went immediately to graduate school, I would never have another opportunity to do something other than be a scientist, and I wanted to make a difference in the world.

I didn't really have a choice where I was sent. When you sign up with CUSO, you go where it sends you. But some country in Africa was my first choice, in part because of the sense of adventure in going there. The continent needed chemistry teachers, and chemistry was something I knew how to do. I had an extraordinary two years in which I wasn't under constant pressure to perform and succeed—like a hiatus. While in Sierra Leone I took my teaching very seriously, but it was something that I could do without feeling as pressured as I had felt at high school and the university.

---

1. M. Meselson and F. W. Stahl, "The Replication of DNA in *Escherichia coli.*" *Proceedings of the National Academy of Sciences* 44, no. 7 (1958): 671–682. http://www.pnas .org/content/44/7/671.short.

I came away from that experience feeling as if I took much more from West Africa than I was able to provide to my students as a chemistry teacher. Sierra Leone was newly independent and still trying to sort out what it meant to be a young democracy. The students whom I taught had very poor backgrounds in science. Their goal was to take and pass the British O-level chemistry exam. That was almost impossible. Perhaps if I had been a better teacher, I could have helped them achieve this goal, but I was making up for years and years of substandard education.

I came back from Africa knowing that I wanted to be a scientist, which was something I had not been totally sure of when I left university. I also knew that I wanted to teach. I love to teach. Even though I wasn't totally successful at it in Sierra Leone, I knew I was a good teacher, and I wanted teaching to be part of my life.

I also love being a scientist because I get to solve puzzles. I am almost a pathological puzzle solver. I love puzzles of all kinds. I love mysteries. I love tackling a problem, digging deep, and seeing if I can solve it.

After Shirley Tilghman graduated from Queen's University in Kingston, Ontario, in 1968, she went to Sierra Leone with the Canadian University Services Overseas and taught high school chemistry.

A MYSTERIOUS GENE

The puzzle I am probably most well known for solving is identifying in the 1980s a mammalian gene that looked nothing like any gene that had ever been described before. It had properties that suggested it was very important—it had been conserved in mammalian evolution—but it also had properties that suggested it had no function whatsoever.

I received a great deal of gratuitous advice over the years while I studied this gene. Some said I should drop it because it can't be important. Others said I should keep at it. But what kept me working on this gene was the sense of mystery it involved. This mystery had not been solved by any other scientist on the planet. That challenge seemed incredibly exciting to me. It was unlike a problem any other lab was studying. The members of my lab and I eventually figured out what it was and what it did. The gene was H19, and it was the very first long noncoding RNA to be described in biology. This discovery opened up a whole new field of study.

Right after I returned from Sierra Leone, I married a Peace Corps volunteer who hailed from Philadelphia. We returned to his hometown, where I went to graduate school at Temple University and he went to law school.

At Temple, I was Richard Hanson's second graduate student. When I entered graduate school, I thought I was going to be an X-ray crystallographer. That would have been in keeping with my strong chemistry background. But Richard, who was my adviser, was trying to understand how cyclic adenosine monophosphate regulates the genes that are involved in glucose homeostasis. This was at the dawn of molecular biology, when it became possible to study individual genes and messenger RNAs. Richard's field was metabolic regulation, which was not a molecular biology, but he gave me the freedom to use molecular tools to approach the questions he was interested in. I got really excited about doing that.

Richard Hanson was a wonderful adviser. In part, this was because he gave me a great deal of freedom and independence. At the same time, though, he was enormously supportive. It was an unusual combination of support when I needed it and space when I didn't. That approach gave me the self-confidence to believe that I was capable of becoming an independent scientist as opposed to someone who could only follow directions.

Speaking as someone who has advised many graduate students, I know that an adviser needs to tailor her relationship with graduate students depending on what she perceives they need. Some students will flourish only if the adviser is keeping pretty close touch with them on a very regular

As an adviser, I've made the suggestion a couple of times to my students that they change fields, but you have to be 100 percent sure. If I'm uncertain, I don't offer that kind of advice because it can have a serious impact on someone. The way you deliver it is obviously very important. [For] a student, [it] takes self-awareness as well as self-confidence to admit that you are going down the wrong path [and to know that a change is necessary]. Some of the symptoms can be unhappiness or boredom—in my case it was boredom. I just wasn't that excited about the subject I was studying. I was missing the "thrill of the chase," which I remembered from my science studies in high school, when I thought that chemistry was the most amazing way to solve puzzles.

basis. But some students will bristle under intense, regular attention to everything they are doing.

Richard sensed in me someone who wanted to test her ability to design a good experiment—I didn't want good experiments handed to me on a platter. Even then, I knew that the most important test of a good scientist is whether he or she can design a great experiment.

My definition of a great experiment is that it answers the question that you are posing, no matter what the result of the experiment. And that describes the Messelson-Stahl experiment. No matter the mechanism by which DNA is replicated, that experiment was going to figure it out. If DNA replicates one way, you will get a certain outcome; if it replicates another way, you will get a different outcome; and so on. I call that a Type A experiment—the design of the experiment contains a test of all the possible options that you can image. Type A's are really hard to design, but if you can create one, you never have to redesign an experiment simply because the results are inconclusive.

Later on, when I started working with mice, it became even more important to create surgical Type A experiments. It generally takes multiple generations of mice to observe results, and since the generation time of mice is really long, and if the experiment isn't well designed, you can discover after 18 months that you have left something out. As a consequence, you might have not have learned anything from a year and a half of work. So when you work on mice, you really need to design great experiments.

I had a wonderful time during graduate school in Richard Hanson's lab. I finished my PhD in three and a half years and published four or five papers. Putting that in perspective, I was working at a time when it was possible to

Shirley Tilghman (*center left*) attended a meeting at Cold Spring Harbor, New York, sponsored by the US National Institutes of Health and the US Department of Energy in 1989. Participants discussed the possibility of sequencing all three billion base pairs of the human genome. The Human Genome Project was begun the following year.

graduate in less than four years and to publish that many papers. We live in very different times now. I have no complaints about graduate school or about having Richard as an adviser.

FINDING A MENTOR

By the time I was heading toward the end of graduate school, I knew that I needed to train under a card-carrying molecular biologist. Richard never pretended to be one, and as supportive as he was, I knew that if I were seriously going to become a molecular biologist, I would need one for a mentor. This was in 1975, at the dawn of the recombinant DNA era, when the first genes were being cloned. It was clear to me this was the future of molecular biology.

The person who really inspired me was Phil Leder. When he came to Temple to give a seminar, I was mesmerized! I found his work on globin genes to be fascinating, and Phil was such an articulate speaker. So I applied to be a postdoc in his lab. It was a minor miracle that he hired me. At the time, some National Institutes of Health fellowships were reserved for foreign scientists, and as a Canadian I qualified. Phil took me on, and it was the best thing that ever happened to me, with the exception of the birth of my children. Phil changed my life.

Phil knew how to isolate and study RNA populations, which is what molecular biologists were doing at that time. But he didn't know how to clone genes—that was my project. When I started at the lab, my job was to figure out how to clone the mouse beta globin gene. At the time, no one knew how to clone a single-copy mammalian gene.

Two of us were given that assignment. The other postdoc was David Tiemeier. It was very lucky that we really liked each other and had good personal chemistry because giving two postdocs the same project could have been a recipe for disaster. But we divided up the jobs that needed to be done and worked well together. It was the most exciting time. We were able to prove that mammalian genes were organized discontinuously in the mammalian genome. It was a huge discovery and propelled both of us into our independent scientific careers.

HANDLING ADVERSITY AND STAYING POSITIVE

I was one of very few women in my university science classes, and none of the professors was a woman. As an undergraduate and graduate student, I encountered male professors who were highly discriminatory. In retrospect, I realize their comments rolled off my back. I believe I was able to ignore because my parents, who were wonderful human beings, gave me a strong ego and confidence in my ability. That is what kept me from feeling as though I shouldn't aspire to be a scientist or that I should take seriously the discriminatory comments made to me along the way.

I remember that when a physics professor at Queen's University made a sexist remark—it was really nasty—I thought to myself, "This guy is such a jerk!" rather than "Maybe he's right: girls shouldn't be physicists."

The greatest gift my parents gave me was sense that no one can tell me what I can do with my life. For that, I thank them every day.

I can't really take credit for choosing such good mentors—by gut instinct or luck I chose two incredibly supportive mentors who never doubted that I was going to succeed in science. Both of them believed that I should aspire to be as good a scientist as I knew how to be. As a result, the incidents that I can recall when someone said in passing, "What do you think you are doing?" did not affect me. My two mentors, whose opinions I respected the most, were telling me whenever appropriate, "You're doing well," "Keep going" and "Don't let anyone tell you differently."

I was really lucky, and I know that my experience was more positive than that of many of my female colleagues. I don't think my experience was representative. I lucked out. And so I worry that when I am quoted as saying didn't face much discrimination, people might conclude that there wasn't *any* discrimination. But I know perfectly well there was.

My advice to young women scientists may sound a little harsh. I tell them not to let anyone turn them into a victim. They have to toughen up, believe in themselves, and work a little bit harder than everyone else. But the minute they turn around and let someone induce them to see themselves as a victim, they are cooked. Much of the time, I try to build up the self-confidence of my women students and not let them fall into a trap I have seen so many women fall into: responding to prejudice by thinking, "People are being so mean to me. Woe is me."

My father was a banker. He worked all his life for the Bank of Nova Scotia, and we moved around a lot. Every time he was promoted, we moved. That constant relocation toughened me up because every three or four years I needed to make new friends and adjust to new places. My mother was a housewife and was immensely supportive but didn't really understand what I was doing. My father's attitude was, "Figure out what you want to do and then do it, for goodness sake." My mother never spent a day thinking her life was anything but wonderful—she is 99 and still has a positive attitude. She always says, "I had a great life." I have two sisters, both younger—one is an economist, the other is a pet food salesperson and raises horses, and both are still in Canada.

Shirley Tilghman retired as president of Princeton University in 2013 and returned to teaching in the Department of Molecular Biology there.

In high school, I was on the chemistry and physics track and did a lot of math. But I also adored English literature and considered whether I should be an English major in college instead of science. To this day I read a great deal of English literature and love it.

My experience applying to college was very different from that of most students today. I applied to only one school—Queen's University in Kingston, Ontario. I never doubted that I would get in since I was a very good student, so I didn't have anxiety about admissions. I had not seen the campus until the day I arrived at the beginning of the first year with my suitcase. Ontario was a long way from my home in Winnipeg, Manitoba, so visiting was impossible. My main concern was how I would pay for school.

My parents didn't have enough money to put me and my sisters through college, so it was important that I get one or more scholarships. I paid my way through college with scholarships and as a teaching assistant in freshman chemistry labs for engineers.

During my sophomore and junior years, I worked for a well-known organic chemistry professor named Saul Wolfe, with whom I published a paper. Wolfe was immensely supportive and never questioned whether he should be hiring a woman to be his lab assistant.

During my third year, I studied with a chemistry professor, who said to me, "Shirley, you are a good student and you study hard, but I just don't think you are going to make a great chemist." The minute he said it, I knew he was right. I believe it is one of the kindest things a mentor can do for someone—to be honest about the likelihood of success. He wasn't saying that I wouldn't be a good scientist—he was just pretty sure that chemistry wasn't the right field for me, that it wasn't in my bones to be a chemist. I had a really good male friend who had chemistry in his bones—and he went on to be a very successful organic chemist. I could see the difference. We got the same grades, but he lived and breathed chemistry in a way that I didn't. I had to work hard at it, and I could sense the difference. But then once I found molecular biology, I could sense the difference between my feelings about chemistry and my feelings about this new field. It has been immensely satisfying to dig deeply to solve puzzles in molecular biology for my career.

ABOUT THE SUBJECT

Shirley Tilghman is professor of molecular
biology and public affairs at Princeton University. She was president of Princeton University from 2001 to 2013, the first woman
to hold that position and the second
woman president in the Ivy League. Tilghman has an undergraduate degree in chemistry from Queen's University of Kingston,
Ontario, and a PhD in biochemistry from
Temple University in Philadelphia.

23 COMMUNICATING AND COMPUTING

William H. Press

"Please, Professor Feynman," I requested. "Won't you do me the courtesy of allowing me to finish my answer?"

As a graduate student at the California Institute of Technology (Caltech), I was taking an oral exam in a course taught by the legendary Richard Feynman, who kept interrupting before I could write even half an equation on the board.

"You think I don't know exactly what you are about to say," he responded, rising ponderously from his desk, his body language communicating his characteristic dry sarcasm. "First, you were going

to say [this] and then [this] and then finally [this]. Am I right?" Of course he was right. I got a B in the course.

Feynman and my thesis adviser, Kip Thorne, influenced me greatly. Feynman taught me to chip away at problems and plot my course toward the final solution with clear mileposts along the way. Kip taught me the importance of communicating clearly, saying that if I have not communicated a result in a way that people can understand and use, it is as if I have not solved the problem. These are precepts I teach my own students.

In the course of my career, I have switched fields about every five years. I have worked in several fields of physics and most recently in computational biology. There are people who devote their life to one subject to make it their own, spending decades diving deeply into one area. I have not done that. I have moved around and worked in many different areas. I can, however, see three common threads in my career: computation, communication, and teaching.

William Press grew up in an academic family in Pasadena, California. He enjoyed writing and doing chemistry experiments in his basement. Photo circa 1970.

A SCIENTIFIC FAMILY

I was born into an academic family in Pasadena, California, and found it easy to become a scientist. My father was a seismologist and geophysicist who directed Caltech's Seismological Laboratory. My mother was a schoolteacher and administrator who loved teaching science to kids. I had great public-school teachers. I grew up in an era when women were limited in their career choices, and many smart women became schoolteachers. I attended elementary school only about 10 years after World War II, so there were also veterans who had gone to college on the GI Bill of Rights and saw public-school teaching as an upwardly mobile profession.

As a kid, I had my "lab" in the basement and tried strange, unguided experiments. Around fifth grade, I had the notion that every liquid must consist of something dissolved into water. To demonstrate this, I placed dishes of different types of liquid outdoors and waited for the water in the liquid to evaporate to reveal the "essence" of the liquid. It was a bit like alchemy. My theory failed when cooking oil did not evaporate no matter how long I waited. In a more successful experiment, I collected micrometeorites containing iron by putting out a large sheet of plastic and waiting several days, then using a magnet to separate the micrometeorites from earth dust.

In high school, I enjoyed studying chemistry and placed fourth in a statewide chemistry competition. I had good math teaching and took calculus at the local community college. High school biology was taught at a level that today would be considered laughable. The cell, I learned, consists of a nucleus floating in uniform cytoplasm. It was 1962, and the recently discovered genetic code had not yet made it into high school textbooks. None of the modern biology topics I work on now had yet been discovered.

I also took physics in high school, but the course was terrible—it was taught by a physical education coach who knew nothing about the subject. I stopped going to class and got a poor grade. I was a smart-aleck kid who often broke the rules. Years later, the coach told me he knew that he hadn't succeeded in teaching physics but knew that I would succeed anyway and that he was proud of me. I was lucky to have had a succession of teachers in high school who put up with my attitude and didn't try to beat it out of me.

At that time, Pasadena had two large high schools, de facto segregated (this was before federally mandated busing). There were about 1,500 kids in my high school class. After graduating, about 50 percent of the kids went into trades, about 40 percent went to community college, and about 10 percent went on to a four-year college. I was one of just 15 classmates to go to college outside of California.

My choice of college was strongly influenced by my parents. I applied to Berkeley, Stanford, Harvard, and Yale and chose to go to Harvard. In the early 1960s, Harvard didn't get that many applications from public high schools in California and may have admitted me in part to add geographical diversity to the class. I certainly wasn't admitted for my expected scholarship. I found that out when summoned to the dean's office for a minor infraction. Reading my admissions folder upside down across his desk, I learned that Harvard's statistical prediction was that that I would be a C student.

A DIFFICULT TRANSITION

Harvard in the fall of 1965 was a shock to me in many ways: living away from home, being in New England instead of Southern California, and attending an elite private college instead of a large public school. The difficulty of the courses was also a shock. Prior to college, I thought I would pursue a career as a pure mathematician. At Harvard, I enrolled in pure-math courses, but pure math turned out to be too hard. I loved math courses and understood the concepts, but I was baffled by problem sets that required me to actually prove things.

Physics courses were also much harder than they were in high school. We studied mechanics during first semester, and at first I had trouble with basic concepts. For example, we were taught that a brick sitting on a table is pushed upward by the table with a force equal to the brick's weight. I couldn't understand how the table knew what the brick's weight was! Then something clicked, and I understood that this was an abstraction and that the brick actually sinks into the table a tiny bit until it achieves force balance. It was a wonderful moment—suddenly I could do freshman physics. Physics is full of such abstractions, and becoming a physicist is learning how to juggle them.

> Up to a certain point, I work on a research topic and don't stop until I really understand it. Then there's a point at which I'm done and need to write up the results. That is when I set deadlines. A lot of life is a tension between these two concepts. At some point, it's important to switch from working to quality to working to deadline, and you need to able to do both in order to do good science.

Second semester I really enjoyed learning electricity and magnetism from Edward Purcell. I would wander into his office, and he seemed to have all the time in the world to talk to me. This seemed amazing because he was a Nobel Prize winner teaching a lecture course with many students. Starting with Professor Purcell, I discovered that you could wander the halls of the Physics Department, ask just about anyone what they were working on, and get an answer. At first, they were startled that a beginning student would ask such a question, but then they would recover and enjoy telling all about their research. Ed Purcell was a wonderful mentor for me throughout all four of my undergraduate years. He taught me the idea of working to quality—studying a problem until you really understand it, from beginning to end. I learned that working to quality is a process that can't be rushed.

In contrast to working to quality, at the undergraduate radio station WHRB I learned about working to deadline. I spent more time at the station than studying for courses, working as a studio engineer, programming classical music, and occasionally announcing. I learned that, ready or not, when it is time for a program to go on the air, it goes on the air. Deadline trumps quality, and you had better plan accordingly.

I earned spending money working two afternoons a week for Professor Gerald Holton's research group studying properties of materials compressed to high pressure. The equipment in the lab was antique and had belonged to Nobel Prize winner Percy Bridgman. However, Harvard had recently installed a time-sharing computer, and for the first time the lab was using a computer to analyze data. I became the computer guru for the lab, and my job was to fit curves to the data. I enjoyed the work and especially spending time with the graduate students and postdocs. When I wasn't hanging out at WHRB, my "home" was in the basement physics lab. But the fact was that I wasn't very good at experimental work. The experimenters in the lab advised me that if I wanted to go on in physics, I should become a theorist. They were right, and that is what I did.

Harvard allowed me to fill my undergraduate studies with odd, unrelated courses. I studied Anglo-Saxon literature in Old English, oral folklore, and intensive Russian (none of which I now remember). I took the minimum number of physics courses required for the major. Despite my lack of preparation in physics, I was accepted for graduate school at Caltech.

By the end of my time at Harvard, we were into the Vietnam War years. I was against the war, not least because I was immersed in an undergraduate culture where everyone was against the war. In truth, I was largely apolitical, or maybe I was what used to be called a "Scoop Jackson Democrat," liberal on social issues but tending to be hawkish on national security issues. During senior year, I applied for a fellowship from the Hertz Foundation. Hertz Fellows were required to sign a pledge to support the United States in a time of crisis. I asked interviewer Lowell Wood whether that meant I had to agree with the Vietnam War. Lowell said absolutely not, that the pledge of support was more of an abstract idea. I doubt that the conservative Hertz Board of Directors would have agreed with Lowell's expansive interpretation. But I felt there was a difference between supporting the country (which I did) and supporting the current government's policies (which I didn't), and I accepted the fellowship without moral qualms.

After I was awarded the fellowship, Lowell invited me to learn more about defense science by spending a summer working for him and Edward Teller at the Lawrence Livermore Laboratory (now the Lawrence Livermore National Laboratory,). My assignment at Livermore was an exercise related to nuclear proliferation. Since I did not have a security clearance, Lowell and Teller were interested in seeing how close I could get to designing a workable two-stage nuclear weapon using only unclassified publicly available information. I found the physics of nuclear weapons to be fascinating. Lowell did give me some help by suggesting physics issues to think about, but even with hints I didn't get very far. I shared an office with a professor from Cincinnati named Lou Witten—not yet better known as Ed Witten's father.

ANOTHER DIFFICULT TRANSITION

After the summer at Livermore, I started graduate school at Caltech. The first year of graduate school was academically the hardest year of my life. I was not sufficiently prepared for rigorous graduate physics courses. After my first year at Caltech, I returned to the Livermore Laboratory from time to time, now with a security clearance that allowed me to attend classified lectures and really find out how nuclear weapons are designed. Of course, I had no idea that this schoolwork was preparing me for what 30 years later would be a stint as deputy laboratory director at Los Alamos National Laboratory.

In my second year at Caltech, I chose Kip Thorne as a thesis adviser and worked with him on general-relativity theory. My thesis was about the stability of rotating black holes. I collaborated with fellow graduate student Saul Teukolsky, who is now a distinguished professor at Cornell. Saul is a talented mathematician who can look at complex equations and intuit the form of their solutions. He discovered what are now known as the Teukolsky equations—linearized versions of the Einstein field equations that describe the dynamics of rotating black holes. Saul and I showed that rotating black holes are stable; they don't just explode and release their rotational energy in a burst of gravitational waves. We worked together—Saul would work analytically and I would work numerically to determine whether the solutions of the relativity equations were stable.

When I arrived at Caltech, the idea that every scientist could personally use a computer to solve problems was just beginning. Before then, if you needed something programmed, you found a professional programmer to do it for you. In that sense, computing was very much a "high priest" activity. If you were a programmer, by definition you weren't also a scientist because programming was itself a technical career. But around this time technical advances such as time-sharing computers, programming libraries, and user-friendly computer languages and compilers changed the assumptions. Every scientist began to interact directly with the computer. It turned out that I was good at scientific programming. I found a number of problems in general relativity in which people had gone as far as they could analytically, but now I could put the equations into a computer and learn a whole new set of things.

In addition to being my thesis adviser, Kip Thorne also helped me improve my writing. Kip would mark up my paper drafts with lots of red ink, and he kept asking what exactly was I trying to say. Not telling Kip, I decided to write a science-fiction novel as a writing-improvement exercise. Every night I would come home and put 1,000 words on paper. At first, I struggled, but by the time I had written 30,000 words, I could quickly sit down and bash out a chapter before dinner. I never finished the novel, but I learned a lot about writing—and also about the tension between quality and deadline.

If you are interested in changing fields, you will be fighting a battle to move into new areas, but not to do so superficially. You should collaborate with people who are experts in your new field and pursue problems *they* find interesting. There are physicists who have moved into biology but who work on problems that the biologists think are physics. So they haven't really moved into biology—they are using biological systems to pursue interesting physics questions. There is nothing wrong with that, but it isn't really a change of career.

I now work on computational biology. My students spend half their time in the bio labs with "real" bio graduate students doing real experiments. That's the only way not to work on sterile problems, what I call "over the transom" problems—a data set thrown over the transom in which you are supposed to try to find interesting things. Sometimes "over the transom" can succeed, but it is much better to be intimately involved in the original gathering of the data. Then you understand its limitations and how the data can be made better for sophisticated analyses. I tell my graduate students that they don't have to learn how to pipette but that they should understand exactly what is being accomplished by each stage of pipetting and how it will affect the data that they will eventually be analyzing.

Richard Feynman was my other important influence during graduate school. Several times a week, the graduate students in Kip's group would have lunch with Feynman in the cafeteria, known as The Greasy. Within the Physics Department, Feynman had advocated for a modern relativity group and the hiring of Kip. Even though Feynman was not directly working in gravitation (despite having famously taught a course about it), he continued to track everything that was happening in our group. He taught us to carefully identify the problem we wanted to solve rather than trying to explore a large or vague region. His ideas were applicable to both experiment and theory. For example, if one of the students were to say, "I'm gathering data on XYZ, and when I have enough data, I will be able to understand the underlying principles," Feynman would turn up his nose and say, "That's not a destination; that's just busy work." He would tell us to figure out one specific thing that we wanted to know and measure or calculate it. He taught us to plot our course from point to point, with mileposts along the way.

I finished my PhD in about three years, and then after a brief postdoc at Caltech, I was offered a faculty position at Princeton. I liked the department and my colleagues there, but not the town or the climate. I was at Princeton for two years before taking a position at Harvard. Already by the time of my PhD, I suffered from intellectual wanderlust, working in relativity but also in astrophysics. I continued to learn to do work in different fields. At Princeton, I worked on fluid dynamics. At the Harvard Center for Astrophysics, I left relativity behind and moved into cosmology. I have switched fields about every five years, even when I have not changed academic jobs, so the switch has not always been visible to outsiders.

Through these changes, the common thread in my career has been computing. This common thread led in the mid-1980s to the writing of the *Numerical Recipes* textbook with three of my colleagues. The book gave me the chance to communicate and teach about an area that wasn't the subject of my scientific research, but rather about the numerical tools used in my research. *Numerical Recipes* was a controversial book in its heyday because it wasn't written by mathematicians who proved theorems about the numerical methods but instead by scientists who used the methods. That was both a strength and a weakness. The many books about numerical analysis out there were not that useful to scientists in physics and chemistry; our book was useful. Over its lifetime, in three editions and multiple computer languages, it sold more than 400,000 copies and continues to sell some thousands of copies per year.

ABOUT THE SUBJECT

William H. Press is the Warren J. and Viola M. Raymer Professor of Computer Sciences and Integrative Biology at the University of Texas at Austin. He has also been a professor of astronomy and physics at Harvard, deputy laboratory director at Los Alamos National Laboratory, and vice chair of the President's Council of Advisers on Science and Technology. Press has an undergraduate degree in physics from Harvard and a PhD in physics from Caltech, where he was a Hertz Fellow.

24 ALWAYS EXPERIMENTING

Richard Miles

My interest in electronics came from my older brother, who would probably be considered a terrorist now. He loved to take apart electronic devices and build new ones. His most passionate enterprise was to circumvent the phone company, tapping into their lines in the middle of the night with his home-built phones to make cost-free long-distance calls. I would accompany him on these nocturnal outings, but I managed to avoid capture—he did not. (He smoked and was caught lighting up.) We lived downstairs in a hillside house in Orinda, California, so we had our own private spaces behind our rooms

that backed up into the dirt of the hillside. His space was filled with piles of electronics—mine was a dark room. We also mixed saltpeter, sulfur, and carbon (portions conveniently specified in the *World Book Encyclopedia*) to make explosives that sent tin cans flying in the back yard and loud echoes in the culvert down the hill. I don't think any animals died, but we shook up a few snakes. I was an experimentalist in the making.

In high school, we had a great math teacher, Mr. Hunt, and I loved chemistry (more explosions), although I never forgave Mr. Campbell for his exam question asking what the smallest measure of water was. His answer? A drop! He was also our swimming coach and a wonderful role model. Mr. Hunt gave us a right triangle with red and blue legs and asked the hypotenuse. (The square root of red squared plus blue squared—I think my answer was "yellow.") The mother of one of my classmates, Lance Chilton, put together a Wednesday evening science club that hosted an invited speaker each week. She also arranged for summer internships for those of us in the club. When I look back now, I am awestruck with the list of speakers and the opportunity she afforded us. It was the first time I met Edward Teller—he spoke on the "physics of stars." That was in 1959. We also had John Wheeler from Princeton and many others. Each of us had a project—I recall that mine was to work with Lance's father on building an electron gun. Mr. Chilton was a master machinist at Lawrence Livermore Laboratory, and his home machine shop was remarkable—he was designing a machine to peel pears: a more difficult job, according to him, than a machine to make bombs since each pear was different. Turning a pear into a sphere was not a success. I got only as far as a few pieces for my electron gun but learned about machining.

THE BERKELEY HILLS

I spent the summers of my junior and senior years as an intern at the Lawrence Berkeley Laboratory in the group of Luis Alvarez (pre–Nobel Prize). The Alvarez group was studying the interaction of K-mesons with deuterium. I recall climbing onto the massive liquid-deuterium bubble-chamber structure, its magnet strong enough to twist my glasses, and seeing the computer system that was being used to quantify the particle tracks. It was an IBM 709 with around a kilobyte of memory. The particle tracking was done by hand, clicking a cursor on the particle tracks in the photographic images, which entered the coordinates into programs called "Kick" and "Pang." The programs analyzed the curvature of the tracks to determine the particle charge-to-mass ratios. I spent a good deal of my time farther up the hill at Project Sherwood, which was seeking to produce controlled fusion using a linear plasma configuration confined with magnetic mirrors. In addition, there was a summer program for high school science teachers that I attended. My second summer I got to know some of the students from Berkeley who were working there. I was invited to the fraternity rush parties that fall, which I attended with my beautiful high school girlfriend. One of my most delightful moments was when I was asked to join a fraternity and declined, saying that I was going to Stanford.

UNDERGRADUATE LIFE IN THE 1960S

Stanford was in my blood—my father and his sisters went there, and his father was a professor of psychology there in the 1920s. (Among other projects, my grandfather worked with Coach Glenn "Pop" Warner studying athletes' reaction times.) My parents were married in the Stanford Memorial Church. I also chose Stanford over MIT because Stanford had an overseas program that sent more than half of the student body to campuses in France, Italy, Austria, Germany, and England for six months. For those of us in science and engineering, the overseas courses satisfied our humanities requirements. I matriculated at Stanford in the fall of 1961, majoring in physics. Freshman year was a blast—I got elected social chairman of our freshman dorm (Cedro House) and had great fun organizing "twisting" parties and other social events but paid little attention to course work. One of my dorm mates was Paul Teller, Edward's son. He arranged for his father to come to debate Ira Sandperl about nonviolence and the Vietnam War, which was just heating up. Edward did not relate well to the nonviolence movement after what the Nazis did in Europe and what the Soviets had done to his Hungarian homeland in 1956. Of course, none of us knew what was in store in Vietnam.

I worked at the (Old) Student Union serving what must have been the worst food I have ever tasted—dried-up "fried" chicken—and listening to Edith Piaf, Pete Seeger, and Peter, Paul, and Mary on the jukebox. The New Student Union (Tresidder) opened the next fall, and my job environment morphed from dark booths, cigarette smoke, and plaintive music to bright sunlight, white tables, and the sounds of dishes and silverware crashing together. I recall sitting on the patio one day in late October 1962 and seeing dozens of B-52s from the Strategic Air Command Base in Merced flying over, headed for the Soviet Union with their nuclear bombs. The beautiful blue sky was laced with white vapor trails all heading northwest. That is my indelible memory of the Cuban Missile Crisis—we thought nuclear war was certain as we watched those planes vanish over the horizon.

In the spring of 1963, I left for France to spend six months at the Stanford campus in Tours, chateaux country, and another three months that fall studying French in Paris and hitchhiking and training around Europe. I learned enough French to get around and loved Europe. I was passing through Tours on my way to Spain when Kennedy was assassinated, and I got to hear the French press saying that the US government was about to fall. When Oswald was shot a few days later, I thought they might be right. I have never seen such an outpouring of love for America as I did in the next few weeks. I returned home for Christmas and went back to Stanford, where I switched majors to electrical engineering.

On April 5, 1964, I received a telephone call from my parents telling me that my brother had died in an automobile accident at his school, Pacific University in Forest Grove, Oregon. That was the hardest day of my life. We were very close; we never even once fought. He had been a major contributor to the theater at Pacific, acting, stage managing, and directing. In his honor, the school named the theater the Tom Miles Theater. I now serve on the Board of Trustees of Pacific University in honor of his legacy.

My father was transferred from the Shell Development Research lab in Emeryville, California, to New York City in 1962, and my parents bought a wonderful house in Old Greenwich, Connecticut, on the water with a floating dock. I spent the summers of 1964 and 1965 there, sailing on Long Island Sound and working the first summer for Canada Dry as a quality-control chemist and the next for CBS Laboratories as an electronics technician. I recall crawling over piles of burlap bags filled with ginger root at a warehouse in Brooklyn, taking samples for analysis. Apparently, the ginger root was collected in the villages of Jamaica and sold to Canada Dry by the pound. The villagers sometimes threw other stuff (donkey dung, for example) into the mix to increase the weight, so we needed to determine the percentage of ginger in order to pay accurately for the shipment. At CBS labs, we were developing background sound-canceling microphones, and I was the guinea pig who sat in the high-sound-pressure noise chamber and spoke into the microphones. The sound level simulated the interior of an armored personnel carrier, which could reach 120 decibels.

I was active in campus life my junior and senior years, playing on intramural volleyball and football teams, working for Rams Head, the student theater organization, and serving as president of my eating club, El Toro. I lived off campus, and the club provided meals, intramural sports, social events, and a wonderful group of friends. Following the devastating Eel River floods of the winter of 1964, our club put together a spring work project in Covelo, an Indian reservation and logging town in northern California. We repeated this project for several years, building close relations with that community, tutoring high school students, and doing various projects in town and on the reservation.

I went back to Europe with Jim Geenley the summer of 1966. We bought a red Renault R-8 in Paris and drove to Israel. We kept the back seat empty and picked up hitchhikers all along the way, stayed at youth hostels, slept out in the open, and met wonderful people in every country: France, Italy, Yugoslavia (it was still one country), Greece, Turkey, Syria, Lebanon, Jordan, and, finally, Israel. Those 90 days were the trip of a lifetime.

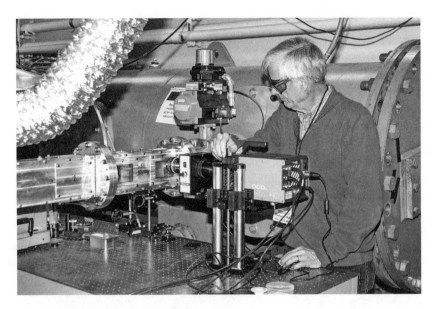

Richard Miles conducted experiments on laser-flow tagging in supersonic nitrogen at the US Air Force Research Lab in 2015.

GRAD SCHOOL OR VIETNAM

Because I had spent time in France in 1963, I graduated late—in January 1966 rather than June 1965. That delay turned out to be a great advantage because I was out of the normal graduate admission sequence. One of my professors, Ralph Smith, asked if I wanted to go to graduate school and said there were research assistantships available in either radio science or lasers. I said yes (Vietnam was the alternative) and began working at the Stanford Systems Techniques Lab on holography and laser propagation under Professor Joseph Goodman.

The Systems Techniques Lab was some distance from the main campus, near Felt Lake, and the work was classified. Due to the distance from the main campus, the interaction with Professor Goodman was not very strong, and the possibility of having a classified thesis was troubling, so after passing the PhD qualification exams (ten 12-minute oral exams in one day), I changed advisers and joined the research group of Professor Hubert Heffner, who was working in quantum systems. I gave up my research assistantship and supported myself as a teaching assistant.

In 1968, I enrolled in a course on satellite design run by Professor Bruce Lusignan. This two-quarter course (Stanford is on the quarter system) had 70

graduate students from the Electrical Engineering, Aeronautical and Astronautical Engineering, Civil Engineering, Operations Research, Industrial
Engineering, and Food Research Departments, as well as from several others.
The goal was to design an earth-resources satellite. We were organized into
subgroups to work on sensors, launch systems, data links, system analysis, and user benefits. We had talks from leaders in the field, took a spring
trip to Vandenberg Air Force Base, Hughes Aircraft, and TRW, and had a
half-dozen other Stanford faculty members associated with the course. Our
satellite was called *Demeter*, for the Greek goddess of the harvest, since the
satellite's main economic impact was predicted to be surveys of crop yields.
I had the honor of leading the final design study and presenting the concept to the International Astronautical Federation in New York in 1968.
After my talk, the rocket technologist Werner von Braun took me aside and
wanted to know about the potential for detecting oil and coal from satellites. All I could think of was the Tom Lehrer song lines, "'Once the rockets
are up, who cares where they come down? / That's not my department,'
says Werner von Braun."

I maintained an active role in campus affairs in graduate school, heading
up the student branch of the Institute of Electrical and Electronics Engineers (IEEE) and the Graduate Student Association. Through the Graduate
Students Association and the IEEE, we sponsored the Technology and the
Third World Conference, featuring John Kenneth Galbraith, who was to
debate Ernest Mandel (a leading Marxist theorist and Trotskyist living in
Belgium) in October 1969. Mandel was refused entry to the United States by
the State Department, however, so we had him speak via recording, and the
debate was instead with Steve Weissman of Berkeley Free Speech Movement
fame. Those were very interesting times, with the Vietnam War protests,
anti–Reserve Officers' Training Corps demonstrations, free speech, co-ed
housing, psychedelic drugs, psychodramas, the Grateful Dead, and so on.
One of my first papers, entitled "The Engineer as a Radical," was published
in the *IEEE Journal*. It was about radical changes created by engineers rather
than about politics, but some did not bother to read it and jumped to other
conclusions.

A COMMITMENT TO SERVE

In 1969, Professor Heffner accepted an appointment as the deputy director
of the Office of Science and Technology Policy in the White House. Before
he left Stanford, he suggested that I apply for the Hertz Foundation Fellowship. I submitted the application the week after I had participated in a

"sociodrama" studying the Stockholm Syndrome. We spent the weekend at a retreat in Big Basin, California, as prisoners, interned for "failing to come to the aid of the country in a national emergency." The Hertz application required a signed commitment to make my skills available to the United States in a time of national emergency. Needless to say, signing that commitment statement gave me great pause—Would I be called up at some point? What is meant by serving the country? At the end of the interview, after I fumbled my way through an atomic physics question, we talked about what commitment to country means. Interviewer Lowell Wood was compelling, explaining the phrase and putting my mind to rest. I feel that I have honored that commitment over the years. For example, I now serve on the US Air Force Scientific Advisory Board, and I thank Lowell for his clear explanation.

Once I had the Hertz Fellowship, I approached Professor Steve Harris and asked if he would serve as my adviser. I had some vague ideas about real-time holography, but he convinced me to work on nonlinear optics. Without the fellowship, I am sure I would not have been able to work in his group because he had a full complement of graduate students.

I began by developing a new optical parametric-oscillator concept to produce coherent infrared light locked to hydrogen fluoride spectral features. That did not work very well and involved handling very corrosive hydrogen fluoride gas. After about a year and an almost completed thesis, Steve suggested we begin looking at third harmonic generation in phase-matched gases. I was not in any hurry to complete my degree since that would end my student deferment and possibly land me in Vietnam, so that became the subject of my thesis. The flexibility I had to address this new topic was also due to the freedom provided by the Hertz Fellowship and the willingness of the manager of the Air Force Office of Scientific Research Program, Howard Schlossberg, to change the focus of the research once it became clear that this was a promising new direction. The fellowship funding was very generous in those days, and it gave me the resources to get a private pilot's license, which became relevant for my future career as a professor of mechanical and aerospace engineering.

LASERS AT PRINCETON

I completed my degree in June 1972 and spent the summer writing a paper based on my thesis before joining what was then the Department of Aerospace and Mechanical Sciences at Princeton University. That department had a wide array of research activities, including propulsion, fluid

Richard and Susan Miles visited the Chama River in New Mexico in 2015.

mechanics, flight dynamics, and materials science as well as its own airport, airplanes, and large-scale research facilities at the James Forrestal Campus. I joined the Gas Dynamics Lab, where there was an interest in gas-dynamic lasers. Nonlinear optics did not seem like a natural match for the department at the time, so I was very pleased that the department had the foresight to bring me on the faculty.

I began by studying carbon monoxide lasers, but my first contribution to gas dynamics was to seed hypersonic helium and supersonic nitrogen with a few parts per million of sodium vapor and do laser-induced fluorescence with a tunable dye laser. The results were great—you could see all the shock structure and tune the laser to highlight the various velocity components. That study started me on the development of both linear and nonlinear spectroscopic methods for flow diagnostics, which has been one of the continuing themes of my research over the past 43 years.

A few years later the of Aerospace and Mechanical Sciences Department's center of gravity shifted to the Engineering Quadrangle, and the department changed its name to Mechanical and Aerospace Engineering. My research now is located primarily at the Engineering Quadrangle, where we use small-scale supersonic wind tunnels; nanosecond, picosecond, and femtosecond lasers; pulsed high-voltage discharges; electron beams; and microwaves to study plasma aerodynamics, plasma-enhanced combustion, advanced laser diagnostics, and stand-off trace-detection concepts.

I have had a very satisfying career at Princeton, serving as the chairman of the engineering physics program for 16 years, director of graduate studies, interim chair for the Department of Mechanical and Aerospace Engineering, and adviser to many incredibly talented graduate and undergraduate students. I have been surrounded and encouraged by wonderful colleagues. I have learned more by teaching that I ever did as a student. I started interviewing for the Hertz Foundation in 1981, and in 1988 Wilson Talley asked me to join the foundation's board. I have continued to interview and serve on that board ever since. Seeing Hertz Fellows pass through graduate school to later make groundbreaking contributions to our nation and our society has been one of the great pleasures of my life.

In 1983, I married Susan McCoy, an ob-gyn in Princeton whom I met after she moved to town and bought a house from one of my friends. We have two children, Tom and Julie, both in their postcollegiate *Odyssey* years. We sold our house in 2008 and moved in next door with my mother, living with her until she passed away at 102 in August 2015. She had come to Princeton with my father in 1995 after he began suffering from dementia and living close to the water in Connecticut became problematic. She

Richard Miles (*left*), members of his research group, and their Ukrainian collaborator, Professor Albina Tropina (*right*) visited the École centrale in Paris in 2015.

played the piano every night, even up to the last night of her life, in honor of my father (who died in 1998), her father (Yale class of 1904 and an author of the Whiffenpoof song, according to her), and various boyfriends—she had a piano piece for each of them, including the Whiffenpoof song in honor of her father and "Good Night, Sweetheart" for my father.

I was elected to the National Academy of Engineering in 2011, a great honor reflecting the efforts of many wonderful students, research staff members, and colleagues who have worked with me and who have made critical contributions to my research. At the end of June 2013, I officially retired as a professor, but I continue to conduct research as a senior scholar and maintain my research group, which is as active as ever. I am still enjoying interviewing for Hertz and serving as chair of the Hertz Fellowship and Program Council. The sports I play have changed from swimming, volleyball, and football to golf and tennis, with an occasional ski trip thrown in.

ABOUT THE AUTHOR

Richard Miles is Distinguished University Professor in Aerospace Engineering at Texas A&M University. He is also professor emeritus and senior scholar in the Mechanical and Aerospace Engineering Department at Princeton University. He holds undergraduate, master's, and PhD degrees in electrical engineering from Stanford University, where he was a Hertz Fellow.

25 A MENTOR AND A DREAM

Rainer Weiss

For a young scientist, it is critical to have an influential mentor who believes in you. Without such a person during your early career, it is as if you are swimming alone. Jerrold Zacharias was just such a mentor and became like a second father to me. He believed in me, which helped me believe in myself. Jerrold was essential to my career. After I flunked out of MIT, he helped me complete my degree and get into graduate school. When I was finishing grad school, he arranged for me to get my first job and later helped me get tenure.

Jerrold is also responsible for my career-long interest in gravitation—he planted the gravitation "bug" in my head when I was working in his lab as a technician. Together we built two cesium-based atomic clocks. The first one became a commercial product and the frequency standard in laboratories around the world. The second one, had it worked, would have been so accurate one could have measured the effect of altitude on time, thereby testing Einstein's general theory of relativity.

When I returned to MIT as a professor, I continued to think about experimental measurements of gravity. In a class on gravitation in 1971, I asked the students to calculate whether gravitational waves could be measured using a laser interferometer. The signals in such a system are extremely small and can be easily masked by system noise. After the class, I spent a few months calculating the size of these noise sources and wrote up my notes in a quarterly progress report. In this progress report on detecting gravitational waves with an interferometer, I assumed it would take about

10 years for us to detect such waves. But I was completely wrong. It turned out that it would take another 44 years before the Laser Interferometric Gravitational-Wave Observatory would detect gravitational waves in 2015, and validate the concept we discussed in the early 1970s.

CLASSICAL MUSIC AND ELECTRONICS

I was born to a half-Jewish family in Berlin, Germany, in 1932. My father was a medical doctor, a Communist, and Jewish. My mother was an actress and a Protestant. Our family fled the Nazis, moving first to Prague and then luckily to the United States in 1938. We settled in New York City in 1939. My family and our immigrant friends loved classical music, which, together with my scientific work, has been an important part of my life.

> I worked on gravitational waves all these years because it was fun. It was a pleasure to work on it. There were some less-pleasant parts—the bookkeeping was not fun, nor some of the committee work. But when I got to the lab, it was fun—every minute of it. And the people I was working with were wonderful.

After World War II, much of the electronic equipment coming back from the war became surplus and ended up on Cortlandt Street in Lower Manhattan. I became very interested in electronics and started building high-fidelity vacuum-tube radios in my room using the surplus equipment. From a movie theater in Brooklyn that burned down, I salvaged beautiful, large movie loudspeakers, which I reconditioned. Family friends who came over to listen to a New York Philharmonic concert were astonished by the quality of the sound: it made them feel as if they were sitting in the concert hall. Friends asked me to make high-fidelity radio sets for them, so I started a word-of-mouth business making them. I wasn't planning to go to college.

Although the sound quality from radio broadcasts was of high quality at that time, the sound quality of recorded music was poor. Phonograph records were noisy with a hissing sound, and the hissing bothered me, especially when listening to quiet passages in classical music. I wanted to design a system to change the bandwidth of the amplifier to eliminate the hissing when the music was quiet. But I reached the limits of what I knew. Without understanding more mathematics or filter theory, I couldn't make such a system work. So, to learn how to make better noise-reduction systems, I decided I needed to go to college.

Professor Jerrold Zacharias (*left*) helped Rainer Weiss find his way as an undergraduate at MIT, taught him how to perform precise physics experiments, and sponsored him for jobs. Rai visited Jerrold at Jerrold's summer house on Tobey Island, Massachusetts, around 1975.

I was admitted to MIT and enrolled in the Electrical Engineering Department. In 1950, MIT had a standard curriculum with much less choice than today. During sophomore year, electrical engineering students were taught power engineering—generators and transmission lines. There was no mention of electronics, circuit theory, or the other topics that interested me. By midyear, I was fed up with the required electrical engineering courses and transferred to the Physics Department, which seemed to have fewer requirements and a more flexible approach.

A ROMANTIC DETOUR

During the summer after sophomore year, I worked at a company that made blood-cell-counting equipment. At the end of the summer, I took a vacation to Nantucket and ran into a beautiful woman on the ferry. She was a music student at Northwestern University in Chicago and a wonderful pianist. I thought she was a goddess and fell madly in love with her. We exchanged letters through junior year, and eventually I decided to go to Chicago to pursue her. But once in Chicago, I found to my dismay that she wanted to end the relationship. I spent too long trying to win her back, not considering how MIT might treat my absence from class. When I got back, I was informed that I had failed out of MIT.

After flunking out, I wandered the campus trying to figure out what to do next. In one of the physics labs, there was an atomic-beam experiment whose electronics were clearly in need of improvement. I offered my assistance. I had accidently walked into the laboratory of Jerrold Zacharias. Joining his lab would have a profound impact on my life and career.

Jerrold Zacharias was an influential physicist and MIT professor who made contributions to physics, US defense, and science education. In the 1930s, Zacharias worked at Columbia University with I. I. Rabi and Norman Ramsey measuring fundamental physical properties such as the spins and magnetic and electric properties of electrons, protons, and other nuclei. During World War II, Zacharias was one of the leaders of the MIT Radiation Laboratory, where most of the radar systems used by the military during the war were developed. He also contributed to the Manhattan Project, leading one of the divisions at Los Alamos National Laboratory. After the war, Zacharias returned to basic research, creating an atomic-beam research laboratory housed in the same building where he had developed military radar. It was into his celebrated laboratory that I had stumbled.

I worked as a technician in Zacharias's atomic-beam laboratory for four years. Our goal was to make extremely accurate atomic clocks. We first worked on a version using the Ramsey oscillatory field method. Our project was successful and led to a commercial clock known as the Atomichron. By the end of that time, I had become a first-class lab technician. I had learned how to machine high-precision mechanical systems, and I had learned more electronics. That was the start of my career.

After the success of the Atomichron, Zacharias told me he was interested in testing Einstein's redshift hypotheses—the idea that a clock in a strong gravitational field runs more slowly than one in a weak field. Testing this idea would require a clock of even greater accuracy. His idea was to direct a beam of atomic cesium upward inside a tall chamber. The slow atoms in the beam would be expected to follow a path much like a ball thrown upward, creating an atomic fountain shape. Because the atoms would be observed for a very long time, the clock would be much more accurate than any existing design.

CONNECTING WITH PHYSICS AND MUSIC

Jerrold and I worked closely together on the fountain clock. It was clear early on that we shared an enthusiasm for solving experimental puzzles in the lab. Over time, I found out that we also shared a love for classical music. I had been learning the piano part to Beethoven's Spring Sonata. One day I was working in the lab and whistling the piano part when from the other room I heard someone whistling the violin part. It was Jerrold, carefully whistling the piece as written, even in the proper key. I couldn't believe it— we kept whistling at each other, and it became a duet. I asked him, "Jerrold, how do you know this piece?" And then he shared with me some details of his history and childhood.

> I love to build equipment, I love to make equipment work, and I enjoy work-ing with students. That I might ultimately detect gravitational waves—that wasn't what kept me going. A new student would arrive, and I would say, "Let's make this system work 10 times better than before." We would try a new trick, the trick would work, and we would go out and celebrate. We were there to have a good time. It may seem simple-minded and not success oriented. Call it what you will. But that was my motivation. I did it because it was fun. My advice to any graduate student is to work on a project that you find fun. And if it isn't fun, go find a different project.

Zacharias came from a remarkably musical family. His mother was a professional violinist from New Orleans, a member of a string quartet. His adopted brother, Beryl Rubinstein, was a concert pianist who later became head of the Cleveland Institute of Music. Jerrold knew the Spring Sonata and pretty much the entire violin and piano repertoire, having heard his mother and older brother play them when he was a child.

It was wonderful working with Jerrold on the fountain clock—we spent nights and weekends in the lab and become very close. In working with him so closely, I became like a member of his family: Jerrold and his wife had two daughters, and I became his son. And so, like a son, I helped Jerrold and his family out when I could.

Jerrold and his wife liked to go to Cape Cod to vacation or a month in the summer. In 1954, there was a bad hurricane. Jerrold heard about a house on Tobey Island on Buzzard's Bay not far from the Bourne Bridge, a monstrous mansion from the 1890s, undercut by the hurricane and in danger of falling into the ocean. Jerrold said to me, "I'd like you to go down there and look at the house. Check out the plumbing, and tell me what you think of the structure."

I drove down to the Cape, and what I saw was remarkable. Half of the house was not supported anymore. It was scary to go inside; the house looked as if it were about ready to tumble into the bay. I examined the furnace and the plumbing—it was all in good shape. The wiring looked good; everything seemed up to code. Everything seemed fine except for the fact that the house was about to fall off a cliff.

I went back to Cambridge and said to Jerrold, "Everything looks wonderful, but what are you going to do about the cliff?" Jerrold told me not to worry. That would be his business. He proceeded to bring boulders from New Hampshire, and he filled up the whole cliff with the boulders. The house is still standing, looking across Buzzard's Bay.

Our fountain-clock experiment did not succeed because the slow atoms were scattered out of the beam by the more numerous fast ones. After the fountain clock, I continued to do other experiments as a graduate student in Jerrold's lab. I worked for seven years there and had a terrific time. During my time in graduate school, I got married, and my wife got pregnant. It was the best time of my life—I got to do experiments without having to write proposals and without anyone telling me what to do. That time couldn't last forever, though. I was eventually told that I needed to finish my research and write a thesis. Wanting to wrap things up quickly, I took a new apparatus I had constructed to make an even better clock and measured the dipole moment of hydrogen fluoride. So my official thesis was actually quite boring.

Next I took a job that Jerrold got for me as an instructor at Tufts. I must have done a good job as an instructor because the school made me an assistant professor. I hadn't actually graduated yet at that point, so I was both a professor and still a graduate student at MIT. I taught electromagnetism and a laboratory course. However, I didn't want to stay at Tufts as a teacher. I wanted to work on gravitation, so I started a position as a postdoc with Bob Dicke at Princeton.

A GRAVITY EXPERIMENT AND AN EARTHQUAKE

In Princeton, we designed an experiment to determine if the earth is being excited by a particular type of gravitational wave, which Dicke theorized might exist. We built an apparatus to look for a spherical mode of the earth that would be excited by such waves. Unluckily, just before Easter in March

1964, there was an enormous earthquake in Alaska. The earthquake excited the earth's spherical mode. After the earthquake, the earth vibrated for many years, obscuring any possible measurement of gravitational waves. I enjoyed working with Bob Dicke at Princeton. Because our experiment failed, I was surprised to receive an invitation to come back and teach at MIT as an assistant professor. The invitation was undoubtedly due to Jerrold, who once again was looking out for me.

Upon returning to MIT as a professor, I started a group studying cosmology and gravitation, doing experiments looking for changes in the gravitational constant. I needed to develop new technology such as stabilized lasers and precision mechanical systems, both of which became important for gravitational-wave detection. Our lab included a machine and electronics shop, and we fabricated most of the components for the experiment.

In 1966, the Physics Department asked me to teach a course in general relativity. I didn't know much about the subject—I knew far more about experimental techniques and concepts than about the theory I was planning to test. But, not wanting to miss the opportunity, I lied, saying, "Sure, I can teach relativity." Though I had just signed up to teach a subject I was no expert in, I wasn't too worried. I felt confident that I could figure it out if forced to do so.

I spent a very complicated semester teaching general relativity while learning it. I based the class on experimental tests of relativity because I understood such tests better than the theory. When I eventually had to teach tensor calculus, I managed by learning it a day ahead of the students. In many cases, the students knew more of the subject than I did, but they were kind enough not to make the class unpleasant.

Near the end of the term, we were discussing special topics in the field. One of the topics was Joseph Weber's idea of measuring the excitement of large aluminum cylinders by tensor gravitational waves. Weber had been working on this experiment for several years and was beginning to report results at the time of the course. (Later, in 1969, he announced the discovery of gravitational waves, but subsequent experiments failed to reproduce his results.) The students wanted me to explain his experiment. I knew what he had done, but I didn't understand it. I had just taught the students (and myself) Einstein's theory of gravity, which replaced gravitational forces by a geometric theory of distortions of space and time caused by matter and energy. Weber's experiment was about a dynamic tidal force of gravity pushing on the bar. The experiment bothered me, and I thought about alternatives.

THE ORIGINS OF LIGO

I spent a weekend trying to figure out a way to explain to the students how a gravitational wave might interact with an apparatus that you could build. Suppose you put two free masses in a vacuum, with an atomic clock on each one, and measure the time light takes to go back and forth between them. Then, if a gravitational wave were to come between the two masses, the time required for the light to bounce between them would change. I said, "That's the way you could measure a gravitational wave." This idea eventually became the basis for the Laser Interferometer Gravitational-Wave Observatory (LIGO).

I gave this problem to the students. They were able to do the calculation because they only needed to combine Einstein's special theory of relativity with a wavelike metric from the general theory. Einstein deduced much of his relativity theory using thought experiments. I figured my idea for a gravitational-wave interferometer was also a thought experiment because you wouldn't actually be able to build it with sufficient sensitivity.

Rainer Weiss created precision experimental apparatuses in his MIT laboratory, circa 1978.

A TENURE PROBLEM

Meanwhile, Bernie Burke, a radio astronomer who was head of the division on astrophysics in the MIT Physics Department, told me that I hadn't published enough papers and had better start publishing more or I wouldn't get tenure and would get kicked out of MIT (again). Burke's comment made me angry, and my first thought was—*I don't give a damn if they kick me out.* But Burke was correct—our group hadn't published much. We had been working on technology development—for example, we had developed a stabilized argon laser—but our work hadn't yet led to many publishable physics results. Or at least what I considered publishable. My standards for publication were based on advice from Jerrold Zacharias. Jerrold had taught me that a scientist should publish in a journal only fully completed experiments with results. He didn't think it fair to publish ideas about experiments because it would make other researchers beholden to you, forced to give credit to someone who had only an idea for an experiment but had not made it work. Burke made a suggestion that might help with my publication dilemma, though: my group might perform cosmology experiments such as measuring the spectrum of the cosmic microwave background radiation that had just been discovered.

> If I were a graduate student today, I would study cosmology. The field of gravitational-wave physics and astronomy is breaking open—all of a sudden it is becoming a real field. There are two critical problems we have to solve: What is the dark energy, and what is the dark matter? These are critical because they have completely altered the way we look at the universe. The other area that excites me, as it is becoming a real quantitative science, is biophysics. Students working in that area can do some good and help people, and they will more easily get jobs than students in other areas of physics.

After the general-relativity course was over, I became friendly with one of the students. Dirk Muehlner told me he was interested in working with me on the cosmology experiment that Burke had suggested. We discussed measuring the spectral distribution of the cosmic background radiation to see if it had a peak and was a truly a thermal spectrum. Dirk and I wrote a proposal to NASA and were funded to fly a radiometer to the upper atmosphere in a balloon. It took us five years, but we eventually learned the technique of flying high-altitude balloons and doing delicate experiments remotely. In our first flight, we observed the same spectrum as others had measured from the ground at low frequencies and saw an excess near the black-body peak. Above the peak, the spectrum seemed to fall as a thermal spectrum. The excess was worrisome, but we published the results. In subsequent flights, we found an error we had made, and in the final flights we had pretty well established that the spectrum was consistent with a black body having a temperature near 2.7°K.

At about same time, John Mather was working with his mentor Paul Richards at Berkeley, making a spectrometer to measure the spectrum of the cosmic background radiation. At one point, he joined us in Palestine, Texas, to learn about ballooning and the technology we had developed, such as sending liquid helium up to space to cool the detectors. Although John's experiment didn't work initially (subsequent flights of his instrument did work), in the process he invented what became the Cosmic Background Explorer satellite, for which he would win the Nobel Prize in 2006.

John's spectrometer was a better instrument to use to make measurements of the spectrum than our radiometer. Dirk and I then went on to measure the angular distribution of the cosmic background and got into trouble with corrupting radiation by dust in our own galaxy. It takes a long time to finish experiments. My tenure case was coming up, and I had published only two papers in five years. By luck, our balloon results came out just in time. I squeaked by and got tenure.

By 1971, many researchers had copied Weber's experiment and were seeing absolutely nothing. That was terrifying. There was a public showdown between Weber and Richard Garwin, a respected physicist who had carefully repeated the experiment. The controversy hurt Weber's reputation and for many years ruined the field's credibility. That history was a hurdle we had to overcome when we started pushing for LIGO.

A FIRST PROTOTYPE

In 1972, I began looking more closely at the idea of timing light between free masses as a way of making a gravitational-wave detector. I spent the summer calculating all the noise terms and came to the conclusion that you actually could make a gravitational-wave detector based on a Michelson interferometer. I wrote up my conclusions in a Research Laboratory of Electronics progress report, went to the head of the RLE, and asked for funding to build a prototype. He gave me the money, which is quite amazing.

Our first gravitational-wave experiment was only a 1.5-meter prototype, though building it took us a long time.

Once I started a gravitational-wave program, the policies of the MIT Physics Department got in our way. I wanted graduate students to work on the project. However, at the time departmental policy required that the galley proofs of a journal paper published on the thesis topic had to be in the hands of the committee before a student could be granted a PhD. And the paper had to contain a physics result, not an engineering result. I felt queasy putting graduate students on the project because I couldn't be sure there would be physics papers coming out of the work. So for many years I did all my gravitational-wave research using undergraduates. I eventually got around this problem by starting graduate students on the gravitational-wave project and then later switching them to the cosmic background radiation program so they could write their required paper.

In 1973, I lost my funding. During the Vietnam War, the military was told to fund only science that was directly relevant to its core mission. Cosmology and gravitation research didn't qualify. So I wrote a proposal to the National Science Foundation (NSF) describing a new way to measure gravitational waves using an interferometer. The NSF sent the proposal out for review, and US scientists, some of whom were still working on Weber bars, panned the idea. I was lucky that the proposal was also reviewed by Europeans, who took a different view. A group in Munich at the Max Planck Institute liked the idea and wanted to collaborate. They built a prototype and solved a number of the experimental problems.

With the progress being made in Europe, the field was beginning to gain respect again, and we eventually began a project at MIT. We had two graduate students on the project who wanted to do theses trying to detect gravitational waves. On one occasion, for a whole weekend we closed the street outside our building to traffic so trucks and cars wouldn't affect the measurements. One of the students looked for periodic sources of waves, and the other looked for burst sources. The data gathered that weekend, together with all the technology development, became the students' theses.

When the Physics Department asked the graduate students what they had found using the detectors, they said that they had seen no gravitational signal at the detectable limit of the apparatus. Physics theses usually include measurement of a physical phenomenon, not the lack of a signal. Some department members were actually rude to these hard-working students during their thesis defense because they did not have a positive result. This behavior made me angry. I decided that the next time we built an

Rainer Weiss and members of his research group gather in 2016 in the laboratory where large-scale LIGO prototypes are housed.

interferometer, we would make it large enough so you could detect gravitational waves.

A FULL-SCALE SYSTEM

An important transition in the field took place in 1975. Kip Thorne and I
talked about possible experiments at the California Institute of Technology
(Caltech). Based in part on our conversations, Caltech hired Ronald Drever
from Scotland as a new faculty member and invested $3 million to build an
interferometer, a large amount of money that indicated a vote of confidence
in the approach.

To build a full-scale system, we would need to have industrial participa-
tion. I proposed that NSF fund a planning study that would include commercial partners. The design would be based on our experience from three
prototypes, all of which were now working; the Germans had built a 30-meter
version, showing that the system could be scaled; and Caltech had made a
huge investment.

The NSF decided to fund both Caltech's and my study. The study went well, as did continued research at Caltech. In the early 1980s, Kip Thorne and I met and decided that Caltech and MIT should join forces and collaborate. Our study with industrial partners Stone & Webster and Arthur D. Little suggested that a full-scale experiment with proper optics and lasers would cost about $100 million. It then took a few years to convince all the parties involved, but at that point we were on the way to the large-scale LIGO experiment whose results you now read about.

ABOUT THE SUBJECT

Rainer Weiss is a professor of physics emeritus at MIT. He holds undergraduate and PhD degrees in physics from MIT. He was awarded the Nobel Prize for Physics in 2017 for his work on the Laser Interferometer Gravitational-Wave Observatory (LIGO) and for the first direct detection of gravity waves.

THE PUBLIC SERVANTS

26 IMPROVING THE SITUATION

Jennifer Roberts

During graduate school, I encountered significant challenges and, like many PhD students, seriously considered switching fields or finding an alternate career path. Nonetheless, I continued to pursue my degree. I acknowledge the excellent support I received from my network of family and friends—my husband was a strong advocate for me, and my parents were important resources. Moreover, I adapted to grad school by attacking challenges in a way that was empowering. I paid attention to the struggles that other grad students were facing and saw that they were often the same as my own. This observation allowed me to reframe concerns as environmental or structural rather than personal. I also saw how the system was set up to make certain tasks especially difficult for students and worked hard to improve the graduate experience for future students, an exercise that helped me overcome my own self-doubt.

MATH, SCIENCE, AND MUSICAL THEATER

I grew up in Maryland near Baltimore. Throughout middle school and high school, I enjoyed math and science and found both disciplines to be fun and intuitive. Homework was one engaging puzzle after another. Although I had a number of math and science teachers who encouraged me, it wasn't until much later that I decided to pursue a science-technology-engineering-math career. When I was young, I wanted a career in musical theater. With a love for singing and appreciation for the musicians in my family, I preferred the idea of a career in the arts. My parents were not scientists—my mother worked as a nurse, and my father worked in finance—and both appreciated the importance of the arts. But my parents and teachers encouraged me to pursue science as well because it seemed to come so naturally to me. For college, I attended the University of Maryland at College Park. It had great programs in music and theater as well as in electrical engineering. I ultimately decided to major in electrical engineering and minor in voice.

During freshman year, I took a programming class using the C language, and I loved it. That summer I got an internship at the Johns Hopkins Applied Physics Lab doing government contract research, where I continued to develop my programming skills. My supervisor was a wonderful mentor who would pose research problems for me to work on. One such project was a program to automatically compute circuit-board layouts. I also worked on a biomedical project to reverse engineer dolphins' ears to design an ecolocation system. My supervisor was even working on open-source music software. He created a wonderful learning environment.

During sophomore year, I decided to take more programming because it seemed both useful and fun. During my third and fourth years, I worked on signal processing in Professor Shihab Shamma's neuroscience research lab. We played sounds for ferrets, whose brains react to music the same way human brains do, and we examined responses in their auditory cortex. I also had summer internships at the Naval Research Lab, working on countermeasures in its radar division. After five years at the University of Maryland, I received dual degrees in electrical engineering and computer science.

By the end of my time at College Park, I was seriously dating my future husband. We met in a ballroom-dance class. We started competing in more than a dozen dance styles and ultimately focused on dances such as waltz, tango, and foxtrot.

Having worked on computer science and electrical engineering projects in three research-focused environments, I found the idea of a PhD in these areas appealing. Meanwhile, I spoke with family members who worked as music teachers and professional musicians about the pros and cons of a career as a professional performer, including the need to audition regularly for new positions. I finally settled on a career in science and applied to graduate programs. Wanting to stay on the East Coast, I applied to MIT, Georgia Tech, and the University of Maryland at College Park. My future husband and I decided that we wanted to live in Boston, so I selected MIT, and we got married four years later.

A COMPLICATED ENVIRONMENT

When I came to the MIT Department of Electrical Engineering and Computer Science (EECS) for graduate school in 2004, it had confusing policies that often were not clearly communicated to students. EECS was the largest department at MIT, with 800 students, including more than 100 incoming graduate students each year. As in many large institutions, some students were falling through the cracks.

The confusion started even before enrollment. Each spring the EECS Department hosted a visit weekend for accepted students. The main purpose of the weekend was for the new graduate students to interview with research groups in order to find a position and thereby get funding. However, the invitation did not mention that graduate funding depended on visit-weekend interviews. If you knew about the interviews, perhaps because you had a friend at MIT already, then you had an advantage. Students who knew about the interviews would skip out of morning talks to sign up for time slots. By the time the less-informed students learned about the interviews, most or all of the slots would be filled.

One of my undergraduate friends from the University of Maryland had also been accepted at MIT. He was married and had a family. My friend expected to receive a letter describing financial-aid packages with information on teaching or research assistantships. He told me he couldn't afford to go to a visit weekend at a school without having a position, so he did not visit MIT that weekend—not knowing that one of the reasons to attend the visit weekend was to obtain a supported research position.

After the term started, there was more confusion. I spent much of my first year trying to find a research group that would be a good fit for my thesis work. By the middle of the year, two of my friends and I realized that we weren't the only ones who found the environment to be overly isolating. We asked the department for permission to create focus groups to better understand student experiences and improve communication. Our focus groups led to a number of proposals, including the creation of a week-long orientation that culminated in an off-site retreat. I was fortunate enough that the EECS Department, despite the early problems it posed for graduate students, was receptive to student feedback and ideas. Professors Eric Grimson, George Verghese, and Terry Orlando were instrumental in helping us start numerous programs to improve student life. We got permission to create an orientation program, and the orientation events were well received. A survey later showed that the orientation program had had an overwhelmingly positive impact on how the department felt to students in

As president of the MIT EECS Department Graduate Student Association, Jennifer Roberts led efforts to make the department a friendlier place for students. She and association members received a recognition award from department head Eric Grimson in 2006.

the program. They got to know their classmates before they started school. They acclimatized to the institute and formed study groups more quickly. They felt more comfortable.

During our orientation program, students learned practical details such as how to register for classes and how to prepare for a meeting with their advisers. They found out which research groups were better for students on academic career paths versus industry career paths and which questions to ask to help them assess which research groups might complement their interests.

IMPROVING THE DEPARTMENT

After creating a successful orientation program, I was elected president of the EECS Department Graduate Student Association, and I continued to work to improve the department and make it a friendlier place for students. One of our efforts sought to understand why female students were leaving the department without degrees at slightly higher rates than male students. The department came to us with a request to look into this issue and identify actions that might improve female student retention.

Several professors were curious about whether female students left early because they wanted to start a family. However, family considerations were not a primary factor because most graduate students did not have children, and many were not yet in a position to start a family. Again, we convened focus groups. We found out that the students who stayed in the department were not necessarily happy. Rather, they were just determined to finish their degree and refused to quit! We also found out some of the reasons they were unhappy. For example, when some students presented their work to their research group, other group members would criticize their work rather than offering constructive suggestions to make the work stronger. After learning more about our student body through focus groups, we recommended additional programs to help make the department more student-friendly. Our focus-group research suggested that a particular systematic challenge would often have a stronger negative effect on female students than on male students. Thus, we developed programs designed for all students, on topics ranging from group dynamics and career options to work–life balance. Our working hypothesis was that making the department more student-friendly would also make it more female-friendly.

We also tried to creatively reframe the issue of work–life balance. Before I came to MIT, the EECS Department had held panels directed at young women students in which women professors with young children would talk about balancing an academic career and a family. Graduate Student Association members, both male and female, believed, however, that such balancing should not be labeled a problem just for women. Both male and female students would benefit from information about how to raise children while pursuing a career. We believed that our generation needed to deal differently with this problem if we were serious about empowering women to succeed professionally. As a result, we organized panel discussions with male and female parents from academia and industry and opened the audience to both men and women. The response was enthusiastic, with more than 100 participants filling a large seminar hall.

I was also fortunate enough to attend a week-long workshop for female graduate students organized by Professor Leslie Kolodziejski, a graduate officer for the MIT EECS Department. The workshop addressed a variety of common challenges that women encounter in professional settings. For example, we used case studies to brainstorm productive responses to challenging scenarios. In one example, we were told, "You are the only woman in a group working on a technical problem. You make a comment, and no one responds. A male colleague repeats what you said, and everyone says, 'What a great idea.' How do you respond?" In this scenario, one woman suggested saying, "Oh yeah, that's what I meant when I made that comment earlier." Such a statement allowed the woman to reclaim ownership of her idea and was a much better response than walking away angry while a male colleague received credit for her suggestion.

For each scenario, about a third of the women would nod their head and say, "Yes, that has happened to me." No one person could navigate all of the scenarios, but in the larger group there was always at least one person who had an excellent suggestion for how to deal with each problem situation. So we learned from one another.

One particularly memorable exercise at the retreat explored stereotypes. The exercise incorporated an activity drawn from the experimental literature. Participants were broken into three groups, and each group received a different set of instructions. Each group of female engineers was asked to work independently and refrain from discussing their instructions with the other groups. One group was asked to use art supplies to create a collage that depicted what it means to be an engineer. Another group was instructed to show what it means to be a male engineer. My group was asked to show what it means to be a female engineer. We shared our collages to compare depictions of an engineer, a male engineer, and a female engineer.

Both the "engineer" and "male engineer" collages featured sports cars and stylized technologies along with positive words such as *awesome* and *smart*. To characterize female engineers, our group's collage showed a woman juggling while balancing on a unicycle. She was blindfolded as she attempted to juggle papers and work and research and hobbies and family and children, all while remaining balanced on the unicycle's single wheel. It was vivid picture of a woman taking on an impossibly difficult number of tasks. The depiction of a female engineer provided a stark contrast to the stereotypical depiction of awesome male engineers. It highlighted perceptions of unattainable societal expectations for female engineers as opposed to perceptions that engineers and male engineers are inherently cool.

We later participated in an exercise that allowed us to explore the work–life balance in a very concrete way. Each person received a set of LEGO bricks and a square LEGO baseplate. Each LEGO color represented a different activity, such as sleep, time with a significant other, time with children, research, publishing, teaching, hobbies, and so on. Only a subset of the LEGOs could fit on the $N \times N$ baseplate at the same time, so each person had to choose what percentage of their time should be allocated to different tasks. The exercise provided a very tangible way to examine which trade-offs might best fit our professional aspirations as well as our personal ones.

In the workshop, Professor Kolodziejski also had a segment on resilience. We learned that resilience is a skill that can be learned as opposed to something that one either does or does not have. We discussed how to respond to negative comments and critiques from students and professors and discussed thought processes that bolster resilience during challenging situations.

Jennifer Roberts (*second row, third from left*) and members of the MIT Graduate Women in Course 6 gathered in 2006 to discuss ways to improve their department. One of their programs was a "Family in Academia" panel featuring both men and women speakers.

I also participated in founding the Resources for Easing Friction and Stress (REFS) peer-support program within the EECS Department. As one of the first-trained REFS counselors, I learned about mediation training to help students communicate in more professional, productive ways. The REFS counselors connected students to resources to help them navigate challenging situations with peers and advisers. The program has grown considerably in the years since it began; there are REFS representatives both within each academic department and throughout the institute.[1]

GRADUATE RESEARCH

At MIT, I first worked on systems to improve cardiovascular monitoring in a group at the Research Laboratory for Electronics. I was coadvised by George Verghese and Thomas Heldt, both of whom were excellent teachers and mentors.

It took me some time to understand the academic approach to solving practical technical problems. I prefer to start by understanding the big picture, identifying the most important problem, and then developing impactful solutions. For example, in cardiovascular monitoring, false alarms are a

1. See the MIT Resources for Easing Friction and Stress website at http://refs.mit.edu.

big problem. False alarms distract nurses, leading them to distrust equipment. When I learned about false alarms, my approach was to research methods to best solve the problem.

In academia, however, research often works in the opposite direction. Research groups aim to apply their particular specialties to whatever issues their techniques might address. George Verghese had a mathematical background, and so our group worked to apply a particular type of mathematics to cardiovascular monitoring. Although this approach never felt natural to me, I came to understand the reasons for it later, when I realized that deep specialization is often required for obtaining research grants as one gains a reputation as an expert in certain types of problems.

After completing a master's thesis in Bayesian methods for cardiovascular monitoring, I decided to do my PhD project in Patrick Winston's group working on computational cognitive science. Patrick had written one of the seminal books on artificial intelligence (AI) and led the MIT AI Lab for many years. He also had a very broad perspective that proved helpful to me. When I gave one of my early presentations to the group in which I had laid out plans for a thesis project, the other students said that my approach was wrong and too ambitious. But Patrick looked at my work and announced to the group, "Jennifer has just defined a research program." I was fortunate that he had a broad enough understanding of the research landscape to recognize the worth of my presentation, even if it was too sweeping for a single student's thesis project.

For my PhD work, I studied "cognitively inspired AI," a name we coined for algorithms that would capture types of human learning that contemporary algorithms could not address. My research focused on learning techniques that combined multiple representations. Experts in physics and math have many different representations and switch back and forth between them to simplify problems. For example, in high-energy physics the Feynman diagram simplifies calculations that were previously intractable. I studied how computers might exploit pairs of representations in order to make solutions easier to identify by using a computationally superior representation to illuminate patterns in a representation that was more easily observed.

The Hertz Foundation supported me through five years of graduate school. In my last year, I won a Google scholarship that paid for me to attend the Grace Hopper Celebration for women in engineering at the Anita Borg Institute. The conference is the world's largest gathering of women technologists, designed to highlight the research and career interests of women in computing. It was amazing. Women from around the world were interested not only in creating cool algorithms or making computation faster but also in using computer science to make the world a better place. The conference focused on mentoring and empowering women at all stages in their careers. I appreciated spending time with so many women interested in having a positive impact.

During and after my time at MIT, I worked on cognitive and computer science programs at two defense contractors. Before graduating, I joined Charles River Analytics, which had a wonderful summer internship program and provided excellent mentoring for graduate students. After graduate school, I joined Aptima, which organized technologists into interdisciplinary teams. After several months at Aptima, I started working on a Defense Advanced Research Projects Agency (DARPA) program, developing data-analysis techniques for large networks. From there, I became a technical coordinator for another DARPA program and then became a DARPA program manager.

At DARPA, I have now started two programs in the areas of data-driven algorithms. The Synergistic Discovery and Design program aims to develop data-driven methods to accelerate scientific discovery and robust design in areas that lack complete models. Such areas include synthetic biology, protein design, and emerging solar materials. I also started a program called Cyber-Hunting at Scale, in which we are developing data-driven methods to detect and characterize cyberthreats in order to protect enterprise networks.

ABOUT THE SUBJECT

Jennifer Roberts currently works as a program manager in the Information Innovation Office of the Defense Advanced Research Projects Agency. She is trained in electrical engineering and computer science and has an undergraduate degree from University of Maryland–College Park and a PhD from MIT, where she was a Hertz Fellow.

27 RECLAIMING MY DREAM

Renee Horton

For as long as I can remember, I have always wanted to be a scientist. My older brother, younger sister, and I used to play with dolls in our house. I remember thinking there was something wrong with Barbie. She clearly needed to be wearing a white lab coat like the one my grandmother had made for me. Back then I didn't know what kind of scientist I wanted to be, just that I was going to be one.

Sometime before middle school, I decided that I was going to be an astronaut. Our family would travel to visit my uncle and take a rest stop at the NASA Stennis welcome center in Mississippi. The welcome center had a Lunar Lander and other spacecraft to play with. I was so excited, thinking, "I could do this; I could be an astronaut and go to outer space!" I wasn't sure how to get there, but looking around at the exhibits, I noticed that many of the astronauts had science degrees.

Once I got to college, I pursued my plan to be an astronaut by applying for the US Air Force Reserve Officers' Training Corps (ROTC) program. But I failed my hearing test, which meant I couldn't be in ROTC. That day, as a 17-year-old freshman, I lost my dream to become an astronaut.

The loss was devastating. I acted recklessly, got pregnant, got married, and dropped out of school.

Ten years later, as a 27-year-old single mom with three kids, I knew it was time to reclaim my dream: go back to school, become a scientist, and change the world.

EARLY LEARNING

I was born and raised in Baton Rouge, Louisiana. My dad was a technician at a chemical plant, and my mom worked as a clerk in a store. Our family valued education. My mom went back to school to get a medical assistant's degree, and my dad went back to get a computer science degree. My aunt is also a computer scientist; another aunt does graphic design; and an uncle and a cousin are pharmacists. So many members of my family had technical training, although none of them was a scientist.

My parents were interested in fostering our imaginations. They bought me a telescope and a chemistry set and helped as I took things apart and tried to put them together. I remember figuring out how to hook up the Christmas tree lights to the radio so they would beat to the music. I also loved math and calculating figures in my head. My mom challenged us to work out the budget.

"You have this much money. What are you going to buy?" There was a reward for whichever one of us estimated the closest to the actual dollar amount. My father would take us to Disney World on vacation almost every year, and I would try to memorize the digits of all the license plates. I would also play number games in my head, such as keeping track of how many cars of each make were on the road.

As a kid, I was a tomboy. I played football and baseball with my big brother and his friends. There was an empty lot behind our house, and we spent a lot of time playing sports there. But I don't think we ever talked about what we wanted to be when we grew up and definitely not about science or engineering.

My favorite subject in middle school was science. I fell in love with science as taught by Mr. Merrill at McKinley Middle Magnet School. Mr. Merrill was a large person who reminded me of a ZZ Top member—he rode a motorbike and liked food. He was always excited to be teaching science. To be in his classroom was always amazing. I knew that the passion for science he displayed was also inside of me. In high school, my physics teacher was OK, and so were my math teachers, but none of them was like Mr. Merrill. When he was teaching earth and physical science, his eyes would light up, and that lit a fire inside of me. Mr. Merrill was excited every day to be a science teacher. Through the rest of my time in school, I never had another teacher who had such an impact on me.

When I was in middle school, my dad encouraged me to be an engineer, and I told him that I really wanted to be a scientist. He said he didn't know any black scientists. He did know some black engineers, so I could definitely become an engineer, but he wasn't sure about becoming a scientist.

At home, I often was in trouble, and I would be sent to my room as punishment. That meant I couldn't watch the family television, which was in the living room. But because of the dictionary there, I couldn't wait to be sent to my room to find and learn new words. I think it was only much later that my parents realized how much I enjoyed being punished and sent to my room. I also liked reading. My dad had subscriptions to *Reader's Digest* and *National Geographic*, and I loved reading both of those magazines.

Renee Horton grew up in Baton Rouge, Louisiana, where her father was a technician at a chemical plant and her mother worked as a clerk in a store.

The Gifted Program in middle school was housed in a mixed, half-black school, but the students in the program itself were predominantly white. So in elementary school and middle school, I was always one of just a couple of black students in the class.

GIFTED AND DIFFERENT

When I got to high school, the Gifted Program was added to the neighborhood school, which was in a predominantly black school zone. Once the program was moved into the neighborhood high school, I was no longer a minority in the school I attended. I was still a minority in the Gifted Program because it was still majority white, but I was a different kind of minority—different from the other black kids. Although I looked like the other black kids at school, I wasn't like them, and they knew I wasn't like them. I was very motivated in school, and my thought process was very different from theirs.

When I was taking precalculus, I was really into it; it looked sexy on paper. I would start with pencil and paper and start to calculate, and pretty soon I was finished. I would come in the next day, and the other students would look at me and ask, "Did you finish all of that?" I would be excited and say, "Yes, I did." And they would comment, "You are such a nerd!"

In the Gifted Program, the homework was always challenging, but what was more challenging was learning how to interact socially. I always felt different in one way or another from the other students in both the program and the high school in general. I was happy to leave high school and go to college. The high school wanted to celebrate my graduating at age 16, but I didn't want that at all. It would only emphasize how different I was from the other students. When I got to college, again I stood out, this time for how young I looked. At 16, I appeared to be about 14. My college peers thought a 14-year-old had shown up to attend classes.

My parents wouldn't allow me to go away to college. They felt more comfortable with my living at home, so I started at Southern University, which was the black school in Baton Rouge, one of the historically black colleges and universities. I enrolled in engineering classes, with plans to be a mechanical engineer or an architect. Although I briefly thought about becoming a marine biologist, my dad again encouraged me to be an engineer. Southern University was known as having one of the best engineering programs in the state. My plan was to get an engineering degree, join the Air Force ROTC, become an officer and a pilot, and then apply for the astronaut program. That was the goal for my life, but it didn't work out that way.

LOSING A DREAM

When I was 17, in the second semester of my freshman year at Southern, I failed a hearing test. I was told I was no longer eligible to enter the Air Force ROTC program. I was shocked. I went through more extensive tests, and the audiologist told me that I had a very significant hearing loss.

In hindsight, I realized that some measures my family took when I was growing up were to compensate for my undiagnosed hearing loss. Accommodations were made without my even knowing about them. For example, I needed speech therapy in elementary school. In addition, I recall that in the Gifted Program we did a lot of work independently. When the teacher would lecture, if I didn't pay attention or if I couldn't understand what she was saying, it was fine as long as I completed my assignments. I still function the same way today—if given material to read, I can comprehend it, often better than if someone explains it to me verbally.

I was completely unprepared to fail the hearing test. I felt devastated—as if my entire life's dream had been taken from me. I didn't have another dream, I didn't know what to do, and I wasn't sure if I still even wanted to pursue even part of the dream. So I acted recklessly and got pregnant at 18. I married my boyfriend and dropped out of school. He was in the military, and we moved to Germany. We were in Germany for three years, and then we moved back to the United States for three years. After two kids and seven years of marriage, though, it became clear that we weren't any good together. We were like two kids who had now grown up and realized we hadn't done our lives right.

By that point, I had become serious about going back to school and getting my degree. I told my husband that we could make it work. I would study while he was in the military, and then when he got out, I could have a career. That would be amazing! But he didn't agree with that plan. He wanted to be the one in charge, the sole support for the family. We decided to split up. I don't regret my marriage. He was the guy I married when I was young, and that part of my life gave me my boys.

The year after separating from my husband, I became involved with another man, and we had a daughter. When the nurse gave her to me, her eyes were open, and she had very serious eyes for a newborn baby—deep, thoughtful, look-through-your-soul kind of eyes. She was looking at me, and I was looking at her, and I knew right then that I absolutely needed to go back to school and change the world for her. I wanted her to have all those opportunities that I hadn't had. I wanted her to have the opportunity to be whatever she wanted to be. I didn't want to ever have to say to her what my

Renee Horton received her PhD in materials science with a concentration in physics in 2011, becoming the first African American to graduate from the University of Alabama in the field.

dad told me—that I wasn't sure if a black person can work in a particular field just because I didn't know any black people who had actually done it. I wanted to be able to give to her the kind of courage I didn't have growing up.

BACK TO SCHOOL

When my daughter turned nine months old, I entered a vocational-training program. First, I got an electronics technician degree. I was very good at the work. The instructor would tell me that I was too advanced and did not belong in the class. I told him I didn't care—I was having so much fun.

After finishing the technician degree, I decided to continue and finish my electrical engineering degree at Louisiana State University (LSU). I started at LSU in January 2000 and moved the kids into family housing—a small two-bedroom apartment. My daughter had one of the bedrooms, the boys shared the master bedroom, and I got the living room. We had a little desk that converted into the dining table. We would eat dinner and close the table. Then the boys would go to their room to do their homework.

My daughter turned two in March, and the boys were eight and ten when I started at LSU. Our situation was very convenient. In the morning, the boys could go through a hole in the fence to get to the local elementary school outside the campus, and my daughter was in daycare close by. After the boys went to school, I could drive her to daycare and then go to classes. The daycare closed at 5:30 p.m., and I had a student job that finished at 5:15 p.m., so I could get in the car and pick her up. The boys would walk home from their after-school program. We all would converge at the house by 5:45 p.m. for dinner.

Although I was studying electrical engineering, I really wanted to study physics. I had always wanted to do research, and the inquisitive part of me kept pushing me in that direction. But with a family to support, it seemed more practical to be studying engineering because I wasn't sure if I could get a job with a physics degree. I chose chip design as my concentration and learned about circuit layouts and routing. I also minored in math and got to take some abstract math courses, which were a joy.

After I graduated with a BS, I decided I would continue on at LSU and get a PhD in electrical engineering. By the time I went back to school, I had delved more deeply into understanding my hearing loss. I learned how to accommodate for the hearing loss, and I had started wearing hearing aids. I was also given accommodations at the university, including a note taker and recordings of the lectures for my classes. Upon getting home, I would read through the notes and listen again to the lectures. I had the kids convinced that the lectures were stories and would tell them, "Listen to this story with me!" The kids would tell me, "He said such and such ...," which would help me understand the lecturer; then I would agree with the kids and say, "Oh, he did say that!"

Because of my hearing problems, one professor at LSU told me I was an idiot and didn't deserve to be in graduate school. And I asked myself, "What am I doing wrong?" I just couldn't understand what he was saying. In doing research to better understand my hearing loss, I found out that the loss is centered in the range of frequencies used in speech, which makes it hard for me to understand people who speak English as a second language. Most of my professors in graduate school were foreign born, so, for me, those teachers sounded like Charlie Brown, only in real life and not in a cartoon TV show.

After one year in electrical engineering graduate school at LSU, I decided to transfer to Southern University and enrolled for a master's degree in physics. Studying physics at Southern was difficult: it was a male-dominated area, and the faculty had no qualms about telling us that women belonged at home and not in the Physics Department. I kept thinking to myself that I did not come back to school for this crap. I just wanted to be a great researcher. I asked myself, "What is the problem with these folks, and why won't they just let me do this?"

Renee Horton met with NASA pioneer and mathematician Katherine Johnson in 2017.

One of the summers during my physics master's study, I went to a two-week program in materials science at the University of Alabama. I had become interested in materials science and decided that if the physics faculty at Southern thought I couldn't become a physicist, then I would do materials science. I packed the kids up that summer and moved into an apartment the University of Alabama had provided in a student dorm. I enrolled the kids at a YMCA day camp while I attended classes for those two weeks.

IN LOVE AGAIN

I found the materials science program to be amazing. After my discouragements at Southern, I was back in love with science again, my eyes twinkling and my ears buzzing. I realized that this was something I could do—that I could find my place in science. The University of Alabama program had concentrations: you could specialize in material characterization or in physical, chemical, or biological aspects of the field. There was also an engineering specialization. I realized that with my background I could bridge some interdisciplinary aspects of materials science.

While I was in the materials science summer program, the Southern Region Education Board called me and offered me a scholarship, saying that

I could be successful in Alabama. I did one more semester at Southern and then in January transferred to the University of Alabama. I decided not to finish my master's at Southern—it was taking so much energy to fight the male-dominated system there, and I had these kids, these little people who were always hungry and needed to eat. So when the University of Alabama welcomed me, I decided to go there. I wanted to become a great researcher, get a PhD, finish school, and get a job.

I don't feel that way now about the choice of staying and fighting versus leaving. At the time, it was so tiring to have to fight the system during the day, trying to overcome all the obstacles, and then go home and be a single mom. But now I tell graduate students, "Stay and fight and make your statement. Make them listen to you. And do it now. It makes a bigger impact on the field if you choose to fight at the beginning. And then you won't have to fight as much at the end."

At the University of Alabama, I started my research in superparamagnetism, depositing thin films. I was studying the superparamagnetic limit. I had a NASA fellowship, so I would work on my professor's research during the semester and go to NASA to work on its projects during the summer. My professor eventually said I needed to make a choice; I couldn't continue doing both. But in fact he made the choice for me, deciding I would no longer be his student. He told me I hadn't been progressing sufficiently. So after almost five years of working in his lab, I was forced out by email.

FIGHTING TO BE INCLUDED

I cried. I was angry and upset. I had spent all those semesters working in this old white professor's lab, fighting to be part of his physics group, and then he woke up one day and decided that I wasn't good enough. He didn't even have the guts to tell me in person.

I called several of my peers and told them what had happened, and they started telling me horror stories of similar behavior by their professors and thesis advisers. My response was, "Those professors are just crazy!" It was comforting for me, though, talking to my friends and finding out that I wasn't the only one with a story like this. The lesson I took from this experience was how important it is to have a support network of friends and mentors. I realized that I had been living on an isolated island, and I needed to have people in my corner—people who would believe in me.

In my second year at the University of Alabama, the school realized that I was the first black to go through the materials science program. My adviser thought that shouldn't be celebrated and considered it an irritant, not something to be acknowledged. I wished I hadn't known I was the first black physicist there because it felt like a huge burden. What would happen if I screwed up—how would that be interpreted? Would it mean that black men can do physics, but not black women? I felt as if I were representing the entire race and gender.

Then I received the Black Engineer of the Year trailblazer award and was interviewed by a newspaper. The reporter called my adviser to ask for a quotation, but he wasn't willing to give a quote; he said he didn't think that my being honored as the first black was that big a deal. So the newspaper found another professor willing to give a quote.

My adviser didn't get it—I walked into the lab every day and saw no one else who looked like me except for the cleaning staff. There was no one else who looked like me in the research lab, in the classroom, or in the cubicles. He couldn't understand what that experience was like, so he didn't think my being in the program was a big deal.

So he let me go in March, and it was awkward finishing out the semester. Our relationship officially ended in May 2010. My program did not give a master's degree; it was PhD or nothing, and I seemed to headed for nothing.

Renee Horton is the Space Launch Systems lead metallic and weld engineer at the NASA Michoud Assembly Facility in New Orleans.

I was scheduled to work at NASA again that summer and made some phone calls. I explained that I was essentially free labor and also had funding to come back to Alabama in the fall. I asked each of the researchers what problem they were working on that could serve as material for publication. I was looking for existing projects I could join and collect data. There were four or five interesting projects. I ended up working on a project that I had already been involved with during a previous summer. For my thesis, I characterized self-reacting friction during welding of dissimilar aluminum alloys. I measured the microhardness, strength, and strain fields of the welds. So I became a welder in an area that NASA cared about but in which there was very little published literature.

A CARING ADVISER

The next positive step was that I got a new adviser in the Aerospace Department. Professor Mark Barkey loved the project, and I loved him. He was a fair person and a very thorough researcher. When I showed up with a complete thesis plan, he was a bit skeptical at first, but then he became really interested in the metallurgy. When I showed him the NASA reports I had written and wanted to combine into a thesis, he said, "That's not how you write a dissertation." Under his direction, I rewrote every single chapter, adding background material and references, and changing the dry bullet-point style of a NASA report into a treatise with full paragraphs in a flowing academic style. Barkey was very patient, and we got the thesis done in about a year. As a white professor for whom science isn't about race, he helped restore my faith in humanity.

I graduated, but without any job projects. My older son now had his own life, and so I moved home with two kids to live with my parents. Both my mom and my stepdad had had knee surgery and could no longer live upstairs. So they built a room for themselves downstairs, and the kids and I lived upstairs. I was about to turn 40 that year. My mother is a comedian, and every morning she would say, "Doc," and I would say "Yes, ma'am." Then she would ask, "Did you find a job?" And when I said no, she would say, "Good—because I have some errands for you to run."

By the time I celebrated my fortieth birthday, I had found a job doing copying at the local Office Depot. I loved working at Office Depot. For the first time in my life, I didn't have the pressures of being superwoman. I could go to work, and nobody but the manager knew I had a PhD. I wasn't an odd duck. I wasn't the supernerd at school, the kid who could calculate numbers in her head. I was the lady who was really good at the copy machine. I showed up to work with a smile, left work with a smile, and always did extra work because I was relieved to be a normal person who could fit in and give my brain a rest from having to work overtime.

Those three months living with my mom and working at Office Depot were special—I spent them figuring out what I really wanted in a job. I was also trying to understand my worth as a black woman with a PhD entering the workforce in 2011. I decided that I wanted a job where I could be happy: happy with the location and in love with who I was and what I was going to be doing from then on.

After three months, I found a job on Craigslist. The position was listed as researcher at a company that made high-strength rope. After working at NASA, I did not feel that it would be especially hard to make rope. In fact, when I got there, the workers told me, "Don't worry, this is not rocket science." That comment seemed pretty ironic. From the second week there, I was miserable. I knew it was not the job for me. I had to leave my family, now located in Lafayette, Louisiana, to travel to the rope company's home office in Washington State, and I was going to have to spend time on oil rigs. I was not a good fit for the company, and I was let go before 90 days were up.

BACK TO NASA

So back I went to Office Depot. I was there for only a couple of weeks when NASA called to offer me a job. I first worked at Marshall Space Flight Center in Huntsville, Alabama, for two years and then took a promotion to the Michoud Assembly Facility in New Orleans. I am now the metallic and weld engineer lead and oversee welding of the Space Launch Rocket, which is the most powerful rocket ever built and is the system that will take us to Mars. Our first mission is scheduled for 2019.

We have had to solve novel tooling issues—no one has ever welded anything 23 feet in diameter. So I oversaw bringing online the largest weld tool in the world. We have seen some metallurgical issues, including weld imperfections that we study using a range of metrology tools, such as scanning electron microscopy and phased-array ultrasound. We do nondestructive testing, including ultrasound on every weld using a giant robotic arm with multiple transducers. This is high-tech stuff that looks like something out of a science-fiction movie. Michoud is an amazing place to work—whenever I walk into our facility, I feel as if I am making history.

I am also currently the president of the National Society of Black Physicists (NSBP). We organize conferences and workshops with the goal of increasing opportunities for African Americans in physics. We are a small organization of about 300 members, half of whom are students.

My involvement in the NSBP started my first year of graduate school graduate school. With help from NSBP, I attended the Second International Conference on Women in Physics, sponsored by the International Union of Pure and Applied Physics. The conference, which focuses on increasing the representation of women in physics, was scheduled to be held in Rio de Janeiro, Brazil, in May 2005, and I wanted to be there.

First, I asked my professor: "If I raise the money, will the university help me attend this conference?" Then I called up the host committee in Brazil to find out how I could attend. They told me I couldn't go because there was already an assigned US delegation. When I found out that the US team had no black representatives on it, I asked the committee in Brazil whether I could put together a team of black physicists to go to the conference if I raised the money, and they said that would be fine.

So as a first-year graduate student, I became the team leader of the planned black delegation to an international conference. I started to invite other physicists to join the delegation and coordinated with Apriel Hodari, who was the NSBP's administrative officer and a board member. But then I got a call from Beverly Hartline, one of the US organizers of the conference, who said that the United States could have only one delegation to the conference. And so the three of us—Apriel, Beverly, and I—had a meeting to straighten things out and merge the two teams so we could go to Brazil.

Both Apriel and Beverly became wonderful mentors for me. Apriel taught me not be afraid to be assertive. She taught me that it was OK to be in charge, to make people listen to you. For those lessons, I am grateful to her. Beverly, a white woman trained as a geophysicist, is as hard as nails and will not take any nonsense. I want to be like her. She has worked in the White House, been a department chair, a dean, and a chancellor. She takes a position for two or three years and then moves on to another one. She taught me that it is OK to conquer and move on.

After the success of the Second International Conference on Women in Physics, the NSBP made me an associate member while I was still a student, which was the first time that had been done. And I became a full member when I graduated. After that, I was elected the youngest president in the organization's history. I also mentor students; I have between eight and ten mentees and try to pass on to them what I have learned from Apriel and Beverly as well as some of the other things I have learned along the way to where I am today.

ABOUT THE SUBJECT

Renee Horton is the Space Launch Systems
lead metallic and weld engineer at the
NASA Michoud Assembly Facility in New
Orleans. She has a BS in electrical engineer-
ing from Louisiana State University and a
PhD in materials science with a concen-
tration in physics from the University of
Alabama. She is also the president of the
National Society of Black Physicists.

28 CHOOSING A BALANCED LIFE

Jami Valentine

Before selecting a particular adviser, I talked to his graduate students. One of his students gave me an honest appraisal of what being in the lab would be like.

"Don't join this lab, it's horrible," he told me. Undeterred, I ignored his warning and joined anyway. As an African American physics student, I wanted the legitimacy of working with a top researcher. I wanted there to never be any doubt whether I deserved to be awarded a PhD or any question about the quality of my work.

Still, that student's warning had merit. Although I ended up having a valuable learning experience in this professor's lab, I found his confrontational style very challenging. Over time, having my work be aggressively questioned crushed my confidence. It was difficult to hear repeatedly that my results could not possibly be correct, especially when they actually were correct. My adviser did a great job of training his students to win technical arguments. By the end of their training, his students project confidence about their scientific conclusions. And by the end of my time in graduate school, I confidently concluded that an academic career was not for me.

My caring undergraduate mentors at Florida A&M University set me on a path to a PhD and a professorship. But instead of continuing on that path after graduate school, I became an examiner at the US Patent and Trademark Office. I love working at the US Patent Office. I continue to learn new scientific ideas, some of them in areas closely related to my graduate research. I enjoy patent law and learning how to apply law to science. I found a science career outside of academia that still allows me to interact with students while affording me a more balanced life.

EARLY OPPORTUNITIES

I grew up in Philadelphia and attended public schools through college. We lived near Temple University, and I benefited from some of Temple's resource programs, especially in music. I was selected for additional opportunities because in the second grade I scored at the ninety-ninth percentile on a standardized achievement test. I remember being upset not to have gotten a 100 on the test until my teacher explained to me how percentiles work and I understood how well I had actually done.

My mother worked for the Philadelphia School District for 39 years and was very supportive of my education. My father was not involved in my life. Our extended family had lived in the same Philadelphia neighborhood for many years. My aunt Kathy, who is brilliant, had won a statewide math competition in the 1960s when she was in elementary school. The principal had wanted her to go to a magnet middle school for gifted kids. He tried to get her accepted, writing letters and banging on doors, but the school would not let her in—we believe because she was black.

I came along about 20 years later, attended the same elementary school as my aunt, studied with many of the same teachers, and had the same principal. When it came time to apply to a selective middle school, the principal was determined, saying, "They didn't let Kathy in, but they are going to accept Jami!"

I was bright, scored well on standardized tests, and had the support of my family and community. These factors, combined with a history in which my aunt had not been afforded all the opportunities of Philadelphia's public-school system, made the difference and enabled me to enroll. The year I went to the Julia R. Masterman Magnet Middle School was the first that kids from my neighborhood's elementary school were able to attend.

They called Masterman a laboratory demonstration school for fifth through ninth graders. Masterman is a nationally ranked school known for innovation and student achievement, with flexible instruction for gifted students. I was interested in science, so I joined the science club and grew strawberries on the roof. We also grew mint and used it to make tea, which I remember smelled great.

But overall I didn't have a happy middle school experience. It was a culture shock to go from a small school in a poor neighborhood where everyone knew me to a large school with students from all over the city, including children of wealthy people and celebrities. For example, Julius Erving (Dr. J) of the Philadelphia 76ers had two sons there—one in my grade and one in the grade ahead of me. I was not one of the popular students; I didn't own the cool clothes or the cool shoes. I also didn't do well academically; elementary school had left me underprepared, with gaps in my knowledge.

Nonetheless, I survived middle school, and my high school experience was quite different in both good and bad ways. Whereas the student body at Masterman had been diverse, the high school I attended was 98 percent African American. Masterman had been well maintained; our high school was not. Our boiler was broken and had never been repaired, so we sat in classes all winter long with our coats on. Our school building had eight stories, but the top two floors couldn't be used because they were condemned. The level of instruction in our underserved vocational-technical public high school was much lower than I was used to. For example, in tenth-grade English class we were assigned a book that our class had read in the fifth grade at Masterman. I had been a mediocre student at Masterman, but once I got to high school, I was a star. I didn't need to take any books home since I was able to do all my homework in class. I earned A's and B's without much effort. Since my classes failed to challenge me or engage me fully, I sought out entertainment in extracurricular activities—I was the manager for several athletic teams, a member of the National Honor Society, and a drummer in the Jazz Band.

Our school had track, cross-country, football, and baseball teams. Since there were no fields at the school, our track team practiced a mile away at the public field. One day when the team was returning to the school from practice, there was a shoot-out near the school, and one of the team members got shot. It was so commonplace at the time for people to get shot that we just changed the student's track team nickname to "Half-a-Butt" since he got shot in the behind.

My physics class was the last block of the day, and I would often skip class. However, on exam day I would always show up and nail the test. My physics teacher at the time, Mr. Rabinowitz, told me if I really applied myself, I could be great at physics. I told him I needed a scholarship in order to go to college. That was the beginning of how I got to Florida A&M University (FAMU). Mr. Rabinowitz was the first of several physics teachers and professors who helped guide me in high school and college and to whom I am grateful. I am still in touch with Mr. Rabinowitz, who is now retired and living in Florida.

I applied to a number of colleges, including FAMU. The FAMU alumni association president came to our school to make a presentation and to meet our valedictorian. After his presentation, I raised my hand and said, "Don't you also want to talk to me?" When he asked why, I said that I had the highest SAT score in our school for the past five years. And even though I wasn't the valedictorian, he agreed to meet with me, too.

A COLLEGE SCHOLARSHIP

The FAMU alumni association provided an annual all-expenses-paid trip for high-achieving students to allow them to visit the campus. As a result of my conversation with the alumni president, I got to ride for free on an Amtrak train going from Philadelphia all the way to Tallahassee, picking up students in Washington, DC, and other cities along the way. When I met with the university president, he offered me a full scholarship, including four summers of research internships at Lawrence Livermore National Laboratory (LLNL). It was a great deal, and I accepted. I was one of only about 10 percent of my high school class who went to college.

In my class at FAMU, there were six students on full scholarships sponsored by LLNL. The summer before freshman year I flew to California—my first time on an airplane—and had my first research experience. My research project was to focus X-ray trajectories by diffracting the rays using an opaque sphere. In later summers at LLNL, I worked on a system to analyze the composition of seawater. The system was designed to be mounted on a ship and to look for particular chemical signatures using X-ray fluorescence.

As a freshman back at FAMU, I took the standard curriculum—physics, chemistry, and math. The school also required freshmen to take a lot of social science classes, more than I thought necessary. During the first term, I had such a wonderful time that I was on the verge of failing physics. Instead of studying, I was spending my time socializing at a club for students with scholarships, playing on a woman's flag-football team, and hanging out with friends. As a result, I had to withdraw from my first physics class, but my scholarship was contingent on my majoring in physics. I wasn't dropped from the scholarship. Instead FAMU provided coaching, mentoring and a stern lecture. I was told that my physics sequence would now be off by a semester, and I could take the introductory physics class again in the spring. The faith and patience shown to me helped make me a loyal alumna and fan of FAMU from that point on.

The 1990s were a special time for Florida A&M University. President Leonard Humphries would visit people's homes and churches to tell them why they should come to FAMU and that there were many scholarships available. FAMU attracted many good students—my class was 90 percent

African American and had 60 or 70 National Merit scholars. At FAMU, it felt comfortable and normal to be a black woman on scholarship studying chemistry or physics. When I saw my peers succeeding and moving on to graduate school or law school, it gave me confidence that I could do so as well. It was a wonderful time to be at the university.

In addition to my summer research at LLNL, I did research during the semesters at the Center for Plasma Science and Technology. We created plasma shock waves in a tube and varied the end conditions to see how they affected the rebounding waves. The work was funded by the US Air Force with the goal of making jet engines more efficient and quieter in the presence of turbulence. My adviser for this work was Professor Joseph Johnson III.

Professor Johnson was an important influence on me and on the FAMU Physics Department. He told our class, "You are not at Harvard, but you are doing the same work and using the same textbooks with the same information. All you have to do is learn it." As the first African American to get a PhD in physics from Yale, he knew what he was talking about. He gave us confidence that when we went out in the world, we would be competent and able to compete. He told us, "Physics is not biased—if you can learn what is in the book, you can take it and be great."

After I left FAMU, it started a master's degree program in physics and then later added a PhD in physics. Professor Johnson was responsible for starting those programs. I recently visited him and his wife, both of whom are now retired and in an assisted-living facility. It was lovely visiting him, cheering him up, and letting him know that his students are out in the world doing physics and remembering him.

Another FAMU professor who had a strong impact on me was Ronald Williams. Although I enjoyed his undergraduate classes, I especially appreciated the help he gave me after I got to graduate school and had challenges passing my qualifying exams. Professor Williams helped me prepare for the exams, and he provided copies of his own qualifying exams from the University of California at Los Angeles so I could practice answering the questions. I also appreciated the sense of community he created in the Physics Department. Every year after the homecoming game, he had a barbecue reception at his house for alumni and students, where the old-timers could meet the recent graduates and current students and give them advice. Professor Williams still holds this reception.

The professors at FAMU told our class that if we were going to be physics majors, we had to continue with our education after college and get a PhD. Otherwise, we would be better off majoring in engineering. Even though I didn't necessarily know that I wanted to get a PhD, I absorbed that message.

Jami Valentine completed her PhD research on spintronic-logic memory
devices in 2006. She was the first African American women to earn a PhD in
physics from Johns Hopkins University.

I read school brochures, took the Graduate Record Examination and applied
to several schools. I was accepted at Brown and Vanderbilt, among other
schools. My professors at FAMU had a meeting to discuss where I should
go. They did not want me to go to a school unless they knew another black
person had been able to get a PhD there. Even now, that is a consideration
that I tell students. Do you want to be the first? Do you want to be the one
who breaks the glass ceiling? Or would you rather go to a place where they
have seen black physicists before and you won't be an oddity or a unicorn?

My professors suggested that I go to graduate school at Vanderbilt
because they had known a couple of black students who had gone there,
they had friends who were professors there, and they could look out for me
if I went there. But instead of Vanderbilt I chose Brown because it was the
highest-ranked school that accepted me and was in the Ivy League.

CHALLENGES AT BROWN

Attending Brown was a great experience, and I made many friends there.
But just as in my transition from middle school to high school, I again
experienced culture shock in going from a historically black college and
university in the South where 90 percent of the students were African
American to an Ivy League university in New England boasting a diverse
and international student population.

Compared to FAMU, the students at Brown were much more confident in their academic abilities. That was especially true for the international students, some of whom had taught physics courses in their home countries. The international students consistently scored higher on exams than the American students. It didn't seem to matter where the American students had been undergraduates, whether at state schools, historically black colleges or universities, or Ivy League schools—when the exam scores were published, the international students were in a higher band, differentiated from the American students.

In addition to academics, another challenge for me at Brown was the social scene. Brown was the first place I had lived where I had to work hard to build a social life or to find a date. In part, the challenge was due to the racial makeup of the school and the size of the graduate program. Most of the grad students at Brown were in the social sciences, and there were only four or five African American students in the physical sciences. We would socialize, and that was fine. But later when I transferred to Johns Hopkins, where 80 percent of the African American graduate students were in the physical sciences, it was a different experience. At Hopkins, I could tell a friend in chemistry or biology that I had been in the lab for two days, and he or she would understand.

Another way Brown was very different from FAMU was that at Brown every student could purchase textbooks on a credit account at the bookstore. At FAMU, if you didn't have cash, you had to photocopy pages or borrow a book. But at Brown, of course you could have the book you needed—you would buy it on credit at the bookstore, and you were expected to pay it off by the time you graduated. This made a tremendous difference in allowing me and other students to do the course work.

I was at Brown for three years, taking courses, doing research on superconducting materials, and trying to pass my general exam, but I ended up leaving with a terminal master's degree because I failed the physics qualifying exam. It was devastating to fail the third and final time, especially because I thought I had done well enough that time and had expected to pass.

In February 1998, I attended a life-saving, energizing, and refreshing conference sponsored by the National Society of Black Physicists (NSBP). I spoke with every university recruiter at the conference. I told them that I was a graduate student at Brown and that although I had failed my qualifying exam, I still planned to become a good scientist. I showed each recruiter strong letters of recommendation my advisers had written for me and asked about the possibility of transferring to his or her school.

TRANSFERRING TO JOHNS HOPKINS

One of the schools recruiting at the conference had a diverse group of students at its table, white, Asian, and black. These graduate students seemed remarkably happy compared to the overly stressed graduate students I knew at Brown. After talking to these student recruiters from Johns Hopkins and a Hopkins professor who was also at the conference, I sent in an application and was accepted.

As a physics graduate student at Johns Hopkins, I did research on novel materials for use in spintronic-logic devices. Spintronic devices are a new type of nonvolatile memory that use magnetic properties of materials to hold data. To make efficient spintronic memories, you need materials whose spin state will not flip while the memory is read. Half-metals are a class of materials that have this property because they have only one spin channel available in their Fermi level. Theorists had predicted that some metals, including gadolinium, would be found to be half-metals, but when we tested a variety of these potential half-metals, we found that most did not have the special properties needed for spintronics. We did identify one half-metal, but unfortunately it maintained its special properties only at temperatures too low to be useful in actual devices.

Our eminent professor was a brilliant scientist, but I found his approach as an adviser so challenging that it was difficult for me to finish a thesis in his lab. He never gave me accolades or even told me I had done a good job. He had the highest of hopes for me, but I found the way he communicated to be disheartening, even crushing. There were times when I was doing an experiment or a calculation and doing it well, and he would show up and ask me a question that would shake my confidence. He would ask me how I knew what I was doing was right, and even tell me that it could not be right. I would assure him that my work was correct. After I went back and checked my references, I found that indeed I was correct. But instead of inspiring in me a renewed sense of confidence, such experiences left me feeling disheartened because I was used to the more supportive learning environments I had growing up and at FAMU.

Except for a few visiting scientists, all the postdocs and graduate students in the lab at Hopkins were Chinese, and our adviser was also from China. I was the lone African American in the lab. The other students shared a cultural understanding. They understood our adviser's confrontational style and didn't take his comments personally, whereas I took all the criticism very personally.

If I questioned his approach, my adviser would say that it's not good enough to be right or to think you are right, you must *know* that you are right. He told me that if I give an academic talk and someone in the audience tells me that I'm not right, and I get shaken by it, then that's not OK. So I must be prepared to deflect any rocks thrown at young academics. But such training turned out to be irrelevant to me because by the end of my time at Johns Hopkins, I had decided that I didn't want an academic career that required me to do research, advise students, and be on countless academic committees.

Jami Valentine is a primary patent examiner at the US Patent and Trademark Office. She is also involved with national efforts to improve diversity in physics and created the website AAWIP.com, which honors the contributions of African American women to physics.

I finished my thesis work, and my adviser said that he wouldn't schedule a thesis defense until I had a job. So I immediately found a job. The US Patent Office was giving eight months of paid training. I figured I could take the job, and then during the eight months I could find a postdoc or figure out what my next step would be.

My first year at the Patent Office felt like a period of recovery from post-traumatic stress disorder caused by graduate school. I remember one day when I was a new examiner, and it was 4:45 p.m. The supervisor told us, "OK, everyone, head out." When I said I was going to stay a bit longer and finish the case I was working on, the supervisor said, "No! You will not. Legally at your pay grade you are not allowed to do voluntary overtime. You must go—go to happy hour." It was refreshing to discover that some people don't work 12- or 16-hour days; in some places, eight hours are considered a workday, and employees are encouraged to have a life outside of work.

A PERFECT JOB

It turned out that working at the Patent Office was perfect for me. I was placed in a group in which about a third of the cases were in an area in which I had done research. I was lucky; other examiners were put in groups outside their specialty, but at the time I joined, there was a tremendous need to examine applications in semiconductor devices. I examine patent applications in the fields of active semiconductor, quantum and nanodevices, and magnetoresistive random-access memory or spintronic memory. I work from home in Florida, and telecommute to my office in Washington, DC. I have a fiancé and a son, so I appreciate the flexible hours. When I exceed my assigned caseload, and I often do, I receive a cash bonus. I love being a patent examiner—it provides a really great quality of life.

Outside of the office, I'm a licensed Zumba instructor and have performed with the Patent Office group as part of Community Day. I volunteer and mentor students, speak at local schools, and participate in "Adopt-a-Physicist," an online program sponsored by the physics honor society Sigma Pi Sigma, where high school classes get to choose a physicist and ask that person questions.

I also network with and support African American women in physics through my activities at the NSBP. This work began when I was at Hopkins after the other two African American physics graduate students in the department left without getting their degrees. I was feeling isolated and made a list of all the African American women with a PhD in physics whom

I had met through NSBP or other conferences. I put the spreadsheet on my webpage, and the NSBP mirrored it on its website.

When I graduated from Johns Hopkins in 2006, the total number of African American women who had received a PhD in physics in the United States was about 30. Currently, there are about 90 African American women with a PhD in physics, astronomy, or astrophysics. There are also many African American graduate students in these fields, and I have started adding their names to the list at the NSBP. Many of these students will be the first African Americans in their department. We let them know they are not alone. We have a private Facebook group where they can connect to others in their position and discuss issues ranging from how to choose an adviser to how to maintain personal relationships. And we have a website: www.aawip.com.

I feel that part of my duty is to share with young physicists that they have more career options than they realize. I was told, "You must go get a PhD, do a postdoc, and become a professor." There are many other career options for a person with a bachelor's degree in physics, including law or medical school. I could have taken my bachelor's degree in physics and started working at the Patent Office 20 years ago, but I didn't realize that was an option. Students don't necessarily have to get a PhD—they can leave their program with a master's and have a wonderful career and a good life.

ABOUT THE SUBJECT

Jami Valentine is a primary patent examiner at the US Patent and Trademark Office. She holds undergraduate and graduate degrees in physics from Florida A&M University, Brown University, and Johns Hopkins University. Valentine was the first African American woman to earn a PhD in physics from Johns Hopkins University.

29 FLEXIBILITY AND CHANGE

Deirdre Olynick

I had been working at the Lawrence Berkeley National Laboratory (LBNL) for eight years when my husband said to me, "You're working more hours than a full-time employee but getting paid only as a part-time one." My husband was correct, but I wanted the flexibility of part-time work so that I could be more involved with my kids. The officially part-time job gave me the freedom to be included in their day-to-day activities and contribute to the community. For instance, I spent time volunteering at their school, which included helping with scientific activities and organizing events  at the school. I was fortunate—my work at the lab was interesting and was conducive to flexibility for a working parent. But even at an academic-type environment such as LBNL, there were limits.

In 2006, my supervisor at LBNL's Center for X-ray Optics at the Molecular Foundry was promoted, creating an opening for me to run the Nanofabrication Lab. I was qualified and an obvious choice, but I could not be considered for this management position and retain my flexibility as a part-time employee. So to ensure I could continue to work in a positive environment, I took the affirmative step of recruiting my new supervisor.

As our children got older, I moved to a full-time position at LBNL. Then I returned to school and earned an MBA. Managing a full-time job and family life while attending business school kept me very busy. One of my goals was to send a message to my colleagues and lab management that I was in this career full force, and I wanted to see change in the workplace and in my life.

I have now made a career change from materials science research to the field of global health as the associate director of a global cancer program at the University of California at San Francisco. Our mission is to lower the cancer burden in developing countries through innovation, collaboration, and education. I am excited to work in a new area, combining my experience as a scientist in a research environment with business skills learned in my MBA program.

BECOMING A SCIENTIST

I went to high school in Wilkesboro, North Carolina, a small town in the mountains. My father lived in the New York metropolitan area, and I got to see him only about three times a year. I grew up with my mother and my stepfather and experienced an unhappy and abusive childhood. I saw excelling in school as my ticket out of that environment. Later, my unhappy childhood experiences made me want to excel as a parent.

My circuitous path to becoming a scientist started with a chemistry class in high school. With the support of an enthusiastic chemistry teacher, I applied to the chemistry program at North Carolina (NC) State University. However, during my first laboratory course, I was impatient with the precision chemistry required and decided it was not the right field for me. I initially considered a major in industrial engineering in order to do engineering and work with people. I did an industrial engineering internship at a Northern Telecom factory, where I monitored the ergonomics of workstations. At the time, the managers didn't yet believe in ergonomic injuries such as carpal tunnel syndrome, but the line workers certainly did. Although I found this work somewhat interesting, I decided I wanted to dive deeper into science.

Returning to NC State for my sophomore year, I took an introductory materials science class with Professor John Russ. I was intrigued by his descriptions of material structure and by the aesthetic nature of materials science. I found it to be a visual subject combining art and science, as when we studied metal grains and classified lattice structures. I decided that materials science was the right field for me and created my own major, with a focus in electronic materials. I performed research at the National Science Foundation–sponsored semiconductor-fabrication research center, studying silicon oxide as a gate-insulating material. For that work, I received a stipend that was sufficient to pay for my tuition, room, and board.

At NC State, I had several excellent mentors. Dennis Maher, a professor affiliated with the National Science Foundation center and a former Bell Labs scientist, was my undergraduate research adviser. Dennis provided feedback on my research and life in general. He encouraged me to seize professional opportunities, including a summer internship at Bell Labs. I was also fortunate to study with Mehmet Ozturk, now director of the Triangle National Lithography Center and NC State's Nanofabrication Facility. Mehmet was one of my favorite teachers and taught me the fundamentals of solid-state materials science.

Although there were some women in technical fields at NC State, we were a small minority. Only about one-fifth of the students majoring in materials science were women, and other technical departments had even fewer. With so few women in the department, I studied and worked with men almost exclusively.

I experienced clear and overt sexism as an undergraduate at NC State. One of my professors would look up girls' skirts. Once when I was waiting to interview for a summer position, I overheard a professor on the phone discussing a prospective student applicant. Wondering whether it was worth his time to meet with her, he asked, "Is she pretty?" Such discriminatory attitudes fueled my desire to excel.

I met my husband at NC State, and we were married while I was still an undergraduate. He was a graduate student in chemistry, and I spent a lot

Deirdre Olynick (*center right*), her mentor Murray Gibson (*center*), and their research group gathered outside their lab at the University of Illinois in 1992.

of time in his lab. I also spent time with my brother, who was a graduate student in aerospace engineering. Based on their examples, I decided that I also wanted to attend graduate school. When I told my father I was planning to get a master's degree, he said I was smart enough to complete a PhD and would be limited without one. I followed his advice and applied to the PhD program in materials science at the University of Illinois.

GRADUATE SCHOOL IN ILLINOIS

I arrived in Urbana with some misgivings after having to walk through a blinding snowstorm on my first day of work. But Illinois was a good choice for both my husband and me. It had one of the top graduate programs in materials science, and my husband had been offered a postdoc with a top professor in the Chemistry Department.

My mentor from my internship at Bell Labs suggested that I work with Professor Murray Gibson. Murray had been head of the Physics Department at Bell Labs and had moved to the University of Illinois about a year before I arrived. Unfortunately, he told me that he couldn't take me as a student unless I brought my own funding. I applied for several fellowships and won a Hertz Fellowship, which made it possible to work with Murray. That was fortunate because he turned out to be an excellent thesis adviser and an understanding mentor. I had observed a range of graduate student–adviser relationships at NC State and knew that for me a positive relationship was critical—as important as my choice of thesis topic. Having not been treated well in childhood, I felt it very important to be treated with understanding and respect in my adulthood.

For my thesis, I studied the role of impurities and contamination on the properties of nanocrystalline copper. I built equipment to create gas-phase crystals and used transmission electron microscopy to study their properties under ultraclean conditions. After two and a half years, my husband completed his postdoc and moved to take a job in the San Francisco Bay Area. I remained in Illinois to continue my graduate work. It was a stressful time in which I worked through feelings of abandonment, a remnant of my childhood. Murray Gibson and I worked out an arrangement in which I could rotate between three-month periods of experimental work in Illinois and periods of analysis at the National Center for Electron Microscopy at LBNL. Being away from my husband was still difficult, but this arrangement made it bearable.

I completed a thesis in four and a half years, which was faster than average for an experimental thesis at Illinois. I was able to accomplish this in part because Murray was trained in the European system, where it is more common to finish in four years, and in part because he had spent his career in a corporate setting and so had a different understanding of results needed for a thesis. But I have wondered if he let me complete my thesis early because I was a pregnant woman living 2,000 miles from my husband. However, having won the Materials Research Society Graduate Student Award and the Hertz Thesis Prize for my research, I concluded that my graduate research was more than sufficient to qualify for a PhD.

Deirdre Olynick received her MBA in 2016. She celebrated afterward with her daughter, Elena, and her husband, Brian.

I chose to have children before starting my professional career because of a medical issue. The doctor said that if I wanted to have children, it would be better to have them sooner rather than later. My husband and I decided that we did want to have children, so I got pregnant in my last year of graduate school. That decision certainly changed the course of my career.

Before my first child, I was very career focused. I was an advocate for women in the scientific workforce and planned to excel at both parenting and working professionally. However, after the birth of my son, Niels, career seemed less important. It was a turning point, and I was thinking about taking a year off before starting in the workforce. Murray encouraged me to line up a job before having the baby. He was concerned that if I opted out of science, I might not opt back in, something that often removes women from the scientific community.

It was tricky interviewing for jobs at the end of graduate school while I was pregnant. I needed to secure a job quickly, before an interviewer could be biased by my obvious pregnancy. I was able to do one on-campus interview before I moved to California, five months pregnant, to write my thesis. After that, I started cold-calling companies and was able to secure several interviews. I was conflicted about whether to tell a potential employer during the interview that I was pregnant, but I decided to tell the interviewer if it interfered with the company's required start date.

CHALLENGES OF WORK AND FAMILY

I joined Applied Materials' New-College-Graduate Program three months after Niels was born. Preparations were not easy. My husband and I had to figure out where to live and to secure daycare. I had to face the realities of being a breast-feeding working mother. Fortunately for me, the job was exciting. I was attending management- and technical-training classes with my cohort in preparation for rotation assignments within the company.

Our house and daycare were close to my workplace but 55 miles from my husband's job at a pharmaceutical company in Richmond, California. We considered living midway between job locations, but that would have required me to commute a full hour with a baby in the car. Our choice left my husband with a two-hour commute, which he found very hard. After a year, I left Applied Materials and found jobs at small companies closer to my husband's workplace. That solved our commuting problem, but I didn't feel challenged at work and continued looking for a better position.

Then I found an Internet listing for a research position at LBNL. I had visited LBNL during graduate school and was familiar with the campus.

Using key words in the job description, I figured out the location where the research was being done. I put on interview clothes, walked inside the building, and asked to meet the supervisor who would be the principal investigator for the position. I had my first interview on the spot and was soon working as a scientist in a part-time position at LBNL.

I was lucky to find a position with interesting work and an understanding manager. I worked in nanofabrication, studying how to fabricate high-resolution features for X-ray microscopy zone plate optics and nanodevices. My manager, Erik Anderson, was supportive of my parenting even though he did not have children of his own. Working part-time allowed me to participate in family and community life. I enrolled my child in a daycare cooperative where I participated in the mornings before heading to work and learned about parenting though an education program and in discussions with other parents.

After five years, my midlevel manager was promoted, and I asked if I could become the lab manager. Because I was working part-time, though, it was not an option. This was one of several times in my career when I had to make a choice between career advancement and time spent with my children and the broader community.

After six years at the lab, I was told I needed to go through a tenure-like process if I wanted a permanent position there. This was a change in policy that caused me considerable stress. I had excelled as a student and was recognized within my research team as a strong contributor. However, I had not prepared for an evaluation that required external visibility. It was discouraging. Despite my strong contributions, I felt that I was not valued by LBNL, and that feeling affected my confidence. Management offered me the option of a lesser title, but I refused and prepared for a full evaluation as a staff scientist. I collected reference letters, pulled together publications, gave academic-level talks, and achieved my goal.

I became a permanent LBNL employee in 2005, although still working part-time, when I transferred to the newly built Molecular Foundry. The foundry was an interdisciplinary environment combining chemistry and physics where I got to do exciting work. My responsibilities included developing my own research program and building the equipment infrastructure. I eventually became a full-time employee in 2011; my children were now older, and I was prepared to increase my engagement in the workplace.

With my transition to full-time professional, I realized there was something missing in my career. In part, I missed having time to spend with the amazing, educated women I met through my kids and community

engagements. Between work, which bled into the weekends, and family responsibilities, I had little time to find the camaraderie of women in
a male-dominated field. In part, I found I could fill this gap by working
on teams. Some of my colleagues at LBNL preferred teamwork, and I was
always happy when I found them. However, because the lab's incentive
system, much like the academic model, rewarded individual contributions,
it was uncommon to work on a large team in our division. I was able to
help create a team, however, in the challenging area of next-generation
semiconductor lithography. I loved that atmosphere—the engaging team

Deirdre Olynick received a thesis prize from John Bravman at the meeting of
the Material Research Society in 1995 for her research on the properties of
nanocrystalline copper.

members, the supportive interactive environment, and the faster progress we are able to make by working together. I also found camaraderie working on organizational issues at the lab, such as the divisional safety committee. In addition, I joined the policy committee for the Women Science and Engineering Council and took a leadership role to improve the climate for women and for workers facing family challenges.

A NEW CAREER

I have recently made a significant career change. Through LBNL's tuition-reimbursement program, I was able to graduate with an MBA from the Wharton School at the University of Pennsylvania. I studied while working full-time and continuing to raise a family. I found going back to school to be a personally transformative experience, but it also sent a signal that I am dedicated to hard work and impactful results.

My original intention was to use this training to facilitate business development and team building at LBNL. However, while at Wharton, I became interested in making impact in health care. I am now associate director of the new global cancer program at the Helen Diller Family Comprehensive Cancer Center at the University of California–San Francisco (UCSF). Cancer is a growing burden on the world's health systems, and more than half the cases are in low- and middle-income countries. Our program is collaborating with institutes in these countries to support clinical and research capacity and to improve cancer outcomes. An important part of our program is career development. We are supporting training for doctors and nurses at both UCSF and institutes in these low- and middle-income countries so they can make global health, in particular global cancer control, a part of their career.

A series of fortuitous events helped me make the switch to a career path I have always admired. When I found the posting for my current position at UCSF, there was clear alignment with my previous experience. The environment at UCSF is very similar to working at a national lab. Both are academic environments, so I understood how to relate to the constituents, and the MBA gave me credibility on the business and operations side of the job. In addition, as I worked on my MBA, I went on two global health trips, one to Cuba and the other to Ethiopia. I also have delivered many scientific talks in other countries and found science to be a great bridge between cultures.

Our kids are now older, which means I have much more flexibility in my life than when they were young. I no longer have to be stressed about being home early. The flexibility is really nice, letting me work longer hours at a new job that I really love and feel passionate about.

ABOUT THE SUBJECT

Deirdre Olynick is a materials scientist with a BS from North Carolina State University and a PhD from the University of Illinois. She was Associate Director of the global cancer program at the University of California–San Francisco. Since the printing of this book, she is the Director of Business Development and Operations at ATOM Sciences in San Francisco, CA.

30 RISKS, LEAPS, AND A SUPPORT NETWORK

Kimberly Budil

"Am I going to be good enough?" was a question I often asked myself in graduate school. My inexperienced and arrogant experimental adviser ignored me for long stretches and then was demeaning in technical meetings. My academic adviser was weak and not helpful. He even had a nervous breakdown! I felt like a fish out of water, at a different school and in a different environment. I wasn't getting any help.

But then the whole situation changed when a fantastic, prominent woman researcher arrived on the scene. With her support, I built up my experiment, got connected to the scientific community, and made more connections to other women scientists. It was more by luck than by design, but I learned about support networks, and my postgraduate experience was far better than my graduate one.

I grew up drawn to challenging activities. In high school, I was first introduced to physics, biology, and chemistry. I was drawn to physics because it is intuitive and fundamental and doesn't require much memorization. With a pocketful of equations, you can derive anything you need. But I also enjoyed fields outside of science. In fact, when I started college at the University of Illinois at Chicago, I intended to be a lawyer but ended up majoring in physics.

I was first exposed to experimental science as an undergraduate while working in the laboratory of Charles Rhodes. It was a large research group, with postdocs and researchers from around the world doing impressive research. I got to take apart and build up lasers, and I wrote a paper about my work. The process made me feel like a real researcher. At the time, it was unusual for undergraduates to have the opportunity to do this type of work. Having a group of students to learn with and being welcomed into the group helped me continue in the field of physics. As a result of that experience, I decided to pursue science as a career and go to graduate school.

During my senior year, I took a graduate course in statistical mechanics. I was the only undergraduate in the class and really struggled with the material. I felt like a colossal failure. However, at the end of the course, the professor, Jim Garland, said he would be glad to write a letter of recommendation for me for graduate school, saying that I was the best undergraduate in the course. He was impressed with my effort and dedication despite my difficulties with the material. His offer made a big impression on me and gave me a confidence boost as I headed to graduate school.

AN EXPERIMENTALIST'S DREAM

For my graduate work, I attended the University of California–Davis in the Department of Applied Science, with my research at the Lawrence Livermore National Laboratory. At the time, I knew very little about the National Laboratories. One of the researchers who would visit our group in Illinois each summer connected me with my adviser. I ended up being this adviser's first graduate student as he started at Livermore. When they flew me out to visit the department and Livermore, I remember being impressed by the amount of resources available to researchers there. It was an experimentalist's dream.

I was excited to join a group that was doing cutting-edge research in an area that was quite interesting to me. However, I quickly became overwhelmed with the research because my adviser expected me to be as competent as he was. He threw me into the deep end to see if I would sink or swim. He ignored me for stretches of time, and I hadn't yet formed a support network to help me work through the difficulties. The biggest push to help me finish my graduate research came when an accomplished researcher from France, Anne L'Huillier, spent a year working in our lab. Without her, I would not have finished my dissertation. Working with her, I started generating the data that were the centerpiece of my thesis. She, along with other researchers in the lab, helped give me the confidence and assertiveness to stand up for myself and own my successes. I was then able to sit down with my adviser and say, "I'm going to finish this, and you're going to help me get there." I also switched to a new academic adviser, Ann Orel, at that time, and she was extremely supportive of me while I completed my dissertation.

Since graduate school, I have learned to select more carefully the types of environments that I am in and the colleagues I work with. I find that I'm drawn to collaborative environments—environments that allow me to interact with enough people that I can find a mentor who is a good fit. Throughout my career, I have developed what I call my "personal board of directors." This is the group of people to whom I reach out when I need help navigating difficult situations—whether managing a conflict, walking the fine line between assertive and aggressive, or, later, when I was thinking about starting a family, balancing personal and research goals.

One thing I noticed in graduate school is that many of the students who bragged about their research early on were ultimately not the most successful. By the point students have gone through an undergraduate physics program, gotten into a graduate program, and passed qualifying exams, they are relatively equal and fairly accomplished. But all egos are not created equal. I have learned to pay less attention to what other people say or how they act and to focus more on my own progress.

After graduate school, I knew that I wanted to stay in a large lab environment. I consider myself a blue-collar scientist—pragmatic and outcome oriented. I never thought I would go into academia. As I was exploring postdoc opportunities, I found a posting from a Livermore group looking for an experimental physicist to study hydrodynamic instabilities using the powerful NOVA laser. The job sounded interesting; I had the experience with lasers and detectors, but I didn't know much about the physics. I talked to the group leaders, who told me I would be a great fit and would learn the physics on the job. This job experience ended up being wonderful, much better than graduate school! I worked on a range of projects and collaborated widely with the goal to publish papers and get a job. My advisers were great—they were committed to helping me learn and develop my career. Working on the big NOVA laser was also fun, and I developed an understanding of the range of work being done at Livermore.

After my postdoc, I transitioned into a permanent research job at Livermore doing experiments with lasers. However, in 1999 Livermore announced that it was shutting down "my" laser in order to build a new, even bigger laser. I figured it would be five years before we had the new laser up and running. I decided to do something different during this gap and transitioned into theory and computational work. I learned how to work with large-scale multiphysics codes and learned more about weapons science. I realized that doing science in an environment that serves national policy merged my interests in science and my original desire to be a lawyer and work for the government. My original career idea was to be more of a social scientist or attorney, but being part of the national security community allowed me to pursue both of my original interests.

SCIENCE AND POLICY

A few years later I was given the opportunity to move to Washington, DC, and help advise the National Nuclear Security Administration, which is part of the US Department of Energy, about its experimental-materials research program. If I had thought about this change extensively, I might not have taken the risk because it meant stepping away from research. I also wasn't sure if my family would come to Washington with me. However, my husband told me, "This is a great opportunity. You should go for it. We will figure it out." And so I went for it! For my first year in Washington, my husband (a "pro" dad) and our two young sons stayed in California. I commuted home every couple of weekends. He and our boys then joined me in Washington for the second year of my assignment. Having a supportive spouse has been immensely important at every step of my career. He takes pride in my successes and supports me in everything I do. I couldn't do what I do without him.

For my first assignment in Washington, I had a huge portfolio of responsibilities. I worried because I definitely was not an expert in all of these areas. I was going to need to do research to support these topics. However, I quickly learned that even though I was not a deep expert, I was one of the most technical people in this environment. I learned how to explain very technical topics to nontechnical people. I was impressed by my coworkers, people who were bright and deeply passionate about the areas in which they worked. They were trying to do the right things and really appreciated my perspective as I explained why certain approaches might or might not work technically. Scientists working in the government can have a big influence, but such work requires patience. It is like trying to turn a cruise ship around; it takes time, but once the ship actually moves, the change is impressive.

During my time in Washington, I learned a number of things about myself. I learned that I am interested in many different things, that I am a strategic thinker, and that I am good at motivating people. Understanding my strengths helped me take a leap and transition into management.

I eventually returned to Livermore and moved into my first serious program-management job, with responsibility for the program that I had been working on in Washington as well as for a broad computational research portfolio. I learned the business of running a large program. Although I missed hands-on research, I learned that I am really good at getting people together and enabling them. The impact factor for a manager is so much greater than for an individual researcher. I learned that I am a good scientist but a great program manager. For a program manager, it is important to be really interested in what people are doing, to understand what they are doing and why, and to give people a purpose in order to motivate them to work in areas of interest. When managing others, I have experienced how rewarding it is for me to watch them make a scientific home run. After five years in that job, I was asked to return to Washington, DC, to serve as a national security adviser to the new undersecretary for science at the Department of Energy, Steve Koonin. This was another offer that was too exciting to refuse.

Despite having worked in Washington for four years, I had always been an employee of the Lawrence Livermore National Lab since graduate school. I began thinking about testing myself in a new environment and looked to try something different. I have recently joined the University of California in my current role as vice president for the National Laboratories. In this role, I get to be a part of the governance and oversight of the laboratories, to work in public service, and to learn about the university and the State of California. One aspect of my work is to identify ways to partner effectively between the National Labs and the rest of the University of California system with the goal of harnessing the 10 campuses, five medical centers, and three National Laboratories together to solve the biggest challenges facing California, the United States, and the world today. We are working on three major initiatives to bring people together: one around high-energy density science, one on understanding the properties of materials, and one on computation in biology, which could revolutionize how drugs are developed. Across the University of California system, ongoing initiatives are focused on climate (including a goal of making the university system entirely carbon neutral by 2025), sustainability of the global food system, and innovation and entrepreneurship.

Throughout my career, I have learned that taking risks has allowed me to discover what drives me and what makes me feel satisfied with my work. Risks, despite being scary at first, are necessary to figure out what you are good at. However, risks often aren't as big as they seem. If you ask yourself, "What is the worst possible outcome?" usually the answer isn't too bad. Be open to opportunities because you never know what you will discover.

ABOUT THE SUBJECT

Kimberly Budil was the vice president for
National Laboratories at the University of
California, Office of the President. Since
the printing of this book, she is the Principal Associate Director of Lawrence Livermore National Laboratory. Budil has a BS
in physics from the University of Illinois–
Chicago and an MS and PhD in applied
science/engineering from the University
of California–Davis, where she was a Hertz
Fellow.

31 PIONEERING GENDER EQUALITY

Wendy R. Cieslak

I hadn't come to Sandia National Laboratories intending to be a pioneer for gender equality at the lab but over time it became clear that if I wanted to reach my full potential as an engineer and a leader, that's precisely what I would have to do. I have never been one to pick my battles, and rather than taking "no" for an answer, I often find obstruction gets me much more fired up.

So when it came time for me to have my first daughter, I persisted through multiple denials and waited out the end of a Sandia president's term to get a part-time policy instituted for new parents. When I had my second daughter, I resisted a manager who took it upon himself to decide what jobs were appropriate for the mother in a two-career household. And every step of the way, I was challenging a lab culture that expected women to be the junior partner in both work life and married life.

The changes I fought for—in both lab policy and culture—I sought for the benefit of myself and my daughters, but I came to realize that these changes were also for the benefit of all women and families in the lab. So as I advanced in my career, I tried more and more to help women by mentoring them in ways I had never been mentored, being the champion or the coach for them when they were obstructed by sexist policies, managers, or peers. Now that I have retired from the lab, this is where I focus my energies—I consider retirement a second career, one where I hope I can help other scientists and engineers reach their fullest potential through mentorship and cultural changes.

EARLY ENCOURAGEMENT

I grew up in Springfield, Massachusetts, where from the beginning my parents and my schools were constantly giving me the message that I could do anything. Perhaps the first indication of this was when the "New Math" curriculum came to our public schools. New Math, or "sets and numbers," as it was called, was a change and a challenge for teachers, but it clicked with me immediately, so I ended up essentially teaching the concepts to my teacher and classmates.

At home, my parents kept me happily oblivious of the gender-gap issue. You see, my mother had wanted to go to college and to work as a nurse or a teacher, but her parents hadn't thought that to be a worthwhile investment, so she worked as a secretary and resented it. Therefore, both my parents always encouraged me, and there was never any indication that "girls can't do that." I remember spending many an hour at my father's corner pharmacy, helping him count out the pills and watching as he mixed together treatments, which perhaps is what first got me interested in science.

My life in Springfield was so successfully isolated from the question of whether girls could be scientists or engineers that the first indication of trouble wasn't until I began thinking about college. A relative who was a professional engineer visited, and he warned me that if I went into engineering, I might face some discrimination—not because I was a woman, but because I was Jewish!

In Springfield's public-school system, the two college-preparatory high schools were called Classical and Technical, the former dedicated more to a liberal arts education and the latter more strictly focused on science, math, and engineering. Though I knew even before high school that I wanted to focus on these technical topics, my mother insisted I attend Classical, mainly because the race riots at the time were worse at Tech. But I ended up getting an excellent scientific education at Classical. The nontechnical classes I took there were perhaps even more important. Since my postsecondary education would end up being focused on science and engineering, all my communications and grammar skills depended on the excellent education I received at Classical. My physics teacher was particularly superb; the textbook he used, *The Fundamentals of Physics* by David Halliday and Robert Resnick, was the same used in first-year physics classes at colleges around the country.

By the time I was a senior, my father had lost his little corner drugstore because he always gave free medicine to those who needed it, so although my teachers wanted me to go to MIT, when Rensselaer Polytechnic Institute (RPI) offered me a full scholarship, I took it. I think my parents were also happy that I would be attending the same school as my brother.

Wendy Cieslak grew up in Springfield, Massachusetts, and attended Classical High School, where she concentrated on nontechnical subjects. She graduated in 1975.

THE BEAUTY OF MATERIALS

RPI had a reputation for being an excellent school for undergraduate education, with accomplished professors teaching the students directly in every class. For example, Professor Robert Resnick taught the introductory physics class using his classic textbook. With strong preparation from high school, I did quite well in my core classes. I was eager to become an engineer because I wanted to use math and science to solve interesting real-world problems, but I didn't know what precisely I would do with my engineering degree or what sort of engineer I wanted to be.

Students at RPI were required to take one class from each discipline of engineering, and I was not excited by the mechanical and electrical engineering courses I took freshman year. Fall of sophomore year, I took an introductory materials course, and it just clicked.

I found it intriguing that something as seemingly static and unmoving as the table in front of me is composed of moving particles, which over extremely long timescales cause the table's structure to change. Making sure structures don't degrade in a catastrophic way is the foundation of many kinds of engineering. I realized that as a materials engineer I would study such questions and get to manipulate the basic building blocks of everything around us.

Dave Duquette, the professor in that introductory class, would always ask one or two students from the class to work in his lab. The job paid well, and I was poor, so I took the opportunity. Because Dave worked in corrosion metallurgy, I started my career working in corrosion metallurgy as well. I worked under two graduate students in two areas of corrosion—stress-corrosion cracking and pitting corrosion. I prepared samples for various tests, cleaning and polishing them until we could examine the crystal grains under a microscope or access the surface with electrodes.

I loved the work in Dave Duquette's lab, but I didn't yet know I would be doing my PhD there. In fact, I didn't know whether I wanted to pursue a PhD at all. Two very different industry experiences at two very different organizations helped determine the future course of my career.

LIFE AS AN ENGINEER

My first industrial experience was a summer internship at a General Electric (GE) plant in Schenectady, New York. I already knew I was interested in metallurgical engineering from my research experience at RPI. Unfortunately, as far as I could tell, my role at GE was to sit at a desk outside my boss's office and play the role of the token female engineer. It was a lousy summer during which I learned nothing technical and suffered sexual harassment. The engineers around me were bachelor's level engineers and didn't seem very interested in their work. By the end of the summer, I knew I didn't want to do what they did, but I didn't know what else I could do.

Because every situation is different, I've never found that there's good one-size-fits-all advice for responding to harassment and discrimination situations. But even in cases where I didn't take any action—as with the manager who threatened retaliation for [my] going to the Equal Employment Opportunity/Affirmative Action Office—I think it was important to keep careful records. That way, if I did feel the need to respond, or if a pattern became clear, I could take forceful and appropriate action.

I went through a phase where my advice to others was to keep your head down since in my case reporting discrimination hadn't been helpful. The lesson I now take from my experience is that it is most important to stand up for what's right. In some cases, I didn't have colleagues or mentors who could vouch for me and corroborate my story, so now I try to play that role for other women. This sort of support is so important since it helps others navigate the process and preempts any attempt at retaliation.

I was fortunate to have some advanced-placement credit stored up, so the co-op office suggested I take a full-semester co-op opportunity at the NASA Lewis Research Center in Cleveland. I had a tremendous experience there! My mentor was studying the mechanics of fatigue and fracture, and he taught me how to set up and analyze his long-running experiments. In our downtime, he would show me how to analyze data and found other ways of teaching me. He once cut a quarter in pieces, revealing the copper metal sandwiched between the cupronickel on the surface. We calculated how the metals at the interface would diffuse into each other and then ran the experiments to check our calculations. I enjoyed everything about that experience—from the experiments to the equations to the quarter. By the end of the internship, I knew I wanted to pursue a PhD in materials science.

I met my future husband, Mike, during sophomore year in a materials science class. We spent months apart while I was in Cleveland, but, as they say, absence makes the heart grow fonder, so when he picked me up after my co-op at NASA, he proposed to me.

With both a fiancé and an interesting lab at RPI, I made the obvious choice to stay there for my PhD. I decided to try to solve an intriguing question that kept coming up in other students' work in the lab: Why is stainless steel that contains a minute amount of molybdenum so much more corrosion resistant than an iron-chromium stainless steel without the additive?

I wanted to graduate at the same time as Mike, who was already two years into his PhD, so it was good that I had a clear research question and a well-established relationship with my adviser. Having a well-defined research direction may have also helped me win a Hertz Fellowship.

My research went smoothly. With financial support from the Hertz Foundation and with the Climax Molybdenum Company supplying all my samples, I had everything I needed to get through my PhD quickly. I used samples with varying levels of both chromium and molybdenum to determine the influence of each element. It was well established that chromium protects steel from corrosion by forming a passivating film on the surface of the metal, but it was unknown how molybdenum further enhances the effect of the chromium. I found that although molybdenum is only negligibly present in the film, it might alter the material structure. I didn't fully elucidate the mechanism of molybdenum's corrosion resistance, which is still something of an open question.

When Mike and I graduated in 1983, there was a downturn in the American steel industry. We were facing the canonical two-body problem. A few corporations around the country might be interested in hiring one ferrous metallurgy engineer, but hiring two was out of the question. We decided

Wendy Cieslak joined Sandia National Laboratories after graduate school and rose through the managerial ranks to the director level. She and her peers gathered outside the Albuquerque, New Mexico, facility in 2008.

to apply to National Laboratories, where we both might find work. I had moved quickly through my PhD so we could graduate together, and we spent time weighing the best way to start our careers if opportunities opened up in a given city. However, as it turned out, the choice was easy. Both Mike and I happily accepted positions as engineers in the materials group at the Sandia National Laboratories' campus in Albuquerque, New Mexico.

OVERCOMING OBSTACLES

My first job at Sandia was exactly what I was hoping for—using math and science to solve interesting, important problems. Though I was a member of the materials group, most of my "customers" (the people coming to me with problems) were members of the battery-research groups. They asked me about ways to minimize parasitic corrosion of the active materials as well as degradation of the materials of construction.

Battery innovation may seem a strange fit for a laboratory whose historical purpose was nuclear weapons design, but all kinds of innovation was—and still is—required to keep our nuclear weapons stockpile safe and reliable. In particular, Sandia National Laboratories had pioneered the concept of the "wooden bomb," a device that could sit static for decades until, heaven forbid, it was needed and in that moment reliably carry out the carefully timed detonation sequence.

Key to this reliability was an extremely stable battery, the so-called thermal battery. The battery was entirely inert at room temperature but contained a solid electrolyte that when heated would melt into a liquid electrolyte that would allow the battery to power the bomb's functions.

Although I wasn't working directly on the thermal batteries, all my work on corrosion was dedicated to ensuring materials would behave reliably after lengths of time sitting in silos. People working on electrical components and batteries would come to me with corrosion-related problems, and I would solve them. These "customers" were happy, I loved my work, and I even discovered a new corrosion-fatigue phenomenon.

But my first supervisor saw things differently. Even though I was solving problems and publishing, achieving just as much as the men in my group, he refused to give me good reviews. When I asked him why, he told me, "You shouldn't care how much money you make; your husband makes plenty."

This was the first instance but not the last of this bias against two-earner households I encountered. Throughout my career, in fact, I found that whenever the person making review and salary decisions could see both my salary and Mike's, mine suffered as a result.

Therefore, also for the first time but not the last, I transferred to get away from a biased supervisor. My colleagues in the battery group knew me and liked my work, so after only a few years I transferred there to work with batteries full-time.

I spent more than seven years in that group and loved it: I was publishing, and my work was getting a good deal of attention at conferences and within the battery industry. I designed a battery that was in prototype production, and my supervisor supported my work.

It was during this time, after becoming pregnant with my first daughter, that I found myself pioneering the lab's policy of part-time work for new

---

One reason I was able to succeed despite adversity was that I had two key people in my life—my father and my husband—who believed in me 100 percent. Another factor was my inability to take "no" for an answer. Sometimes this tendency was my own worst enemy—I definitely made some people angry, and not just the managers who were blocking my advancement. In the end, it was dogged determination that allowed me to change the policies I did and to stay in the field when others may have become too frustrated.

parents. At the time, I was working on three different projects. I realized that I could work part-time to allow me weekday time with my daughter and still put just as much work into one or two of these projects. So I went to human resources, but they said, "No, we don't do part-time."

But once I get an idea in my head, it's really hard to stop me. I contacted person after person in human resources, and all of them said, "No, we don't do part-time." I received this response until I walked into the office of a staffer who, it turned out, had been hoping to start a part-time policy and had been waiting for someone like me.

Because my different projects were separable, and because I told my supervisor that if I couldn't work part-time, I wouldn't work at all (I wanted to take parenting seriously), both this human-resources staff member and my manager were eager to champion my cause. The lab also had a new president who was eager to get the policy working. But he wanted it to be a full policy and not just a one-time exception for me, so I found myself burning vacation time and sick time and every scrap of leave I could while they wrote the policy. The new policy was established just in the nick of time, and I was able to take advantage of it for five years.

Wendy Cieslak and her daughters, Jacqueline (*left*) and Linda, enjoyed the sunshine outside Sandia Laboratories on Mother's Day in 1993.

ENCOUNTERING SEXISM

After more than seven years in the batteries group, I was ready to move up, and my supervisor agreed—in fact, he told me that when it was time for him to move on, he would sponsor me to fill his position and coach me for the interview. But when his promotion came, I was in the second trimester with my second daughter. When I came into his office to congratulate him and ask him for coaching, he laughed and said, "Just look at you! Are you kidding me?" He told me that he and his wife had decided that a family with two young children couldn't handle two professional parents, so he wasn't going to nominate me for the promotion.

I was friendly with this manager and until then had felt strongly supported by him, so when I went to the Equal Employment Opportunity/Affirmative Action (EEO/AA) Office, I told its staff that I just wanted their advice on how to handle the interaction with my manager, but they insisted that they would handle the situation. The next time I saw my manager, it was clear that EEO/AA had called him in for a meeting, and he was very angry. He told me, "I spoke to you as a friend. How dare you?" This was a lesson for me not only about discrimination but also about management—when you're in

Wendy Cieslak retired from Sandia National Laboratories in 2013. She and her husband, Michael, went hiking in 2017.

a position of power over someone, you're never just speaking to that person as a friend. He wasn't just advising me; he had blocked my promotion.

In the end, lab management reorganized the area, reducing the number of departments from four to three. As a result, they didn't have to promote anyone at all. In a twist, I ended up working for the manager who had refused to give me good reviews. At our first meeting, he told me, "If you ever go to the EEO/AA office again, I will see to it that your career at this laboratory is finished."

I did consider reporting him and was keeping records of all our conversations, but at the time I didn't want to risk retaliation. So once again my path upward was blocked until I could find a lateral promotion within the organization. Fortunately, this came within a few years, when I was able to bid on and win a management job in the materials organization.

DRIVING POLICY CHANGE

As the lab grew more confident in the part-time policy I had pioneered, it would eventually allow people in managerial positions to work part-time, but in those early days I had to return to working full-time. I found the job itself to be fantastic—I had 25 researchers in my group, working on everything from metals and corrosion to polymers and adhesion. I loved learning about what everyone was doing and clearing obstacles for them. At that time, managers still had some discretionary funding, and I found that I could make a difference using those limited funds to support projects that advanced key research areas for my staff.

A few years later my husband felt as if he was in somewhat of a professional rut. We were seriously considering leaving Sandia when a professional society offered him the chance to spend a year in Washington, DC, as a Congressional Science Fellow. My management was thrilled because it's hard to get midcareer researchers to take time off to go to DC. They realized that if they could get me to accompany Mike back east, they would have two people who could be effective advisers in DC and then bring back to the lab what they had learned.

There was just one sticking point: for such temporary assignments from the lab, typically the trailing spouse (me) would receive no stipend. But with everything from housing to schooling more expensive in DC, we would be in a much worse financial position than if we stayed in Albuquerque.

When I told my management how much this move would cost, they championed a lab policy change so that trailing spouses in these situations would get a stipend that was a fraction of the leading spouse's. The fraction they arranged for, not coincidentally, matched our increased schooling and daycare costs almost exactly.

This was the second time I catalyzed a lab policy change to help improve my own work–family balance. It was only slowly that I came to see these efforts as for the benefit of women and families in general. Years later another woman called me up to tell me she had finished the job I started—trailing spouse stipends were now full, not fractional.

Our year in Washington was a wonderful experience for me. I managed nationwide grants worth tens of millions of dollars with leading professors in my field and learned how the funding world worked at that level. That year gave me enough confidence and experience that when I returned to my job in Albuquerque, I realized I had outgrown it. I wanted to continue to shape strategy. So when a senior management position opened up in geosciences, I applied for and received the promotion.

I subsequently held various senior management positions over the next nine years. But then, after 25 years at Sandia, it was my turn to feel stifled. I wanted an even more strategic role, and I wanted to return to my first love, materials, so I started casting about for promotions.

None seemed forthcoming at Sandia, but the materials division leader position at Los Alamos National Laboratory opened up in 2008, and I applied. After so many years pushing for changes at work to accommodate my family life, I had to make a significant change to my home life to take this dream job. For three years, I commuted the 90 miles from Albuquerque to Los Alamos, staying in Los Alamos four to five nights a week and returning home on weekends.

Mike was always 100 percent supportive. Throughout my career, he had responded to any obstacle I faced with the comment, "You're a Hertz Fellow. Go out there and do it!" He and my youngest daughter jokingly call this period the "abandonment years" since for much of her time in high school, I was in Los Alamos. On the plus side, I think it gave her some confidence in having her father to herself those years.

After three years, I jumped at the chance to return to Sandia as director for weapons science and technology. At the pinnacle of my career, I was managing projects in so many of the areas I had come to love during my time in New Mexico.

I retired in 2013 after a three-decade career in the National Labs, almost all of it at Sandia. Retirement has opened up a new emphasis for me. I have taken what I learned advocating for equity and work–life balance in Sandia and am using it to help labs and universities as well as the Hertz Foundation to think carefully about how better to provide women and minorities with the mentorship and support they need. So even though I have retired, I'm still using 30 years of experience to make sure the nation's researchers can reach their full potential.

ABOUT THE SUBJECT

Wendy R. Cieslak is a director emeritus at Sandia National Laboratories and a former director of the Fannie & John Hertz Foundation. She previously was the principal program director for nuclear weapons science and technology at Sandia and materials science and technology division leader at Los Alamos National Laboratory. Wendy holds BS and PhD degrees in materials engineering from Rensselaer Polytechnic Institute, where she was a Hertz Fellow.

32 DISAPPOINTMENTS AND SUCCESSES

Ellen Stofan

The NASA Discovery Program provides scientists the opportunity to dig deep into their imaginations and find innovative ways to unlock the mysteries of the solar system. My team and I proposed splashing down a lander into a freezing sea of Saturn's moon Titan, which consists of liquid ethane and methane. The goal was to discover if life exists beyond Earth. Our proposal, Titan Mare Explorer (TiME), was one of three to reach the final stage of the NASA Dis-

covery Program; however, there was funding for only one mission, and the Insight mission was selected to go to Mars.

After being notified that TiME was not selected, I was heartbroken. I was also concerned about the emotions of my wonderful team after informing them of the news. I reminded myself of a Buddhist philosophy that when life is going a certain way, and you think it's bad, it's actually for the good, and when you think it's for the good, it's not. It turned out that way in this case for me as I went on to another challenging job at NASA. Unfortunately, it will soon be dark at the North Pole seas of Titan, and the next window for this type of sea-lander mission will be the mid-2030s. By that time, I will be retired, but I hope to see it happen.

A YOUNG GEOLOGIST

I was born in Oberlin, Ohio, and was always surrounded by role models in science, technology, engineering, and math (STEM). My father was a rocket scientist for NASA at the Lewis (now Glenn) Research Center, and my mother was a science teacher. Due to my father's involvement with rockets,

I was able to travel at the age of four to Kennedy Space Center to see my first unmanned rocket launch. Unfortunately, there was a system failure, and the rocket exploded on the launch pad, which may be the reason I never had a desire to be an astronaut. However, I was always interested in geology, even at an early age. While growing up, I had a huge rock collection and would store different types of rocks all over the house. Even today I still collect rocks—which drives my husband and kids crazy.

[The Jet Propulsion Laboratory] was very supportive of my desire to work part-time when I had young children. They encouraged me to do whatever I needed in order to accommodate my family responsibilities. I was still writing and winning grants and supporting my hours with grant money. So I was productive, which made that offer easier to make. They also gave me the flexibility to return to work full-time upon my family's return from London. NASA is very supportive of employees trying to maintain a work–life balance.

I strongly support these flexible [human-resource] policies. If we are going to have a diverse STEM workforce, we need to accommodate employees' needs as family or personal demands increase. These flexibilities are important, and since it worked out for me, I want it to work out for others.

While I was in elementary school, my mother decided to go back to school to get a master's degree in education. She took a geology course, and the curriculum included a "hands-on" fieldtrip. I was excited to accompany her on that trip because we walked down streambeds with the professor looking at rock exposures. I thought to myself, "Wow, there's an actual career where one can wander around picking up rocks and get paid to do it."

I remember having a stern science teacher named Mr. Scherzer, whom I was slightly afraid of, and I thought I was his least-favorite student. At the end of the school year, we completed a science project, and he talked with me about my future. He mentioned that he had the highest hopes for me to excel and become a great scientist. After I was appointed NASA chief scientist, I received an email from Mr. Scherzer asking if I remembered him—of course I remembered him! People like Mr. Scherzer who had faith in me and expressed it made a tremendous difference in my life.

When I was 12 years old, my father worked on the Viking project, the first US mission to land a spacecraft safely on the surface of Mars. Days leading up to the rocket launch, NASA implemented educational programs and activities for families. Carl Sagan was one of the speakers who explained why we were exploring Mars and briefly described the geology of Mars. After his presentation, I thought, "Wow—geology and NASA and planets—I'm sold." That's when I decided to become a planetary geologist.

I recently had a conversation with a colleague on the experience of being a woman in a STEM field. We talked about the additional stresses that women and minorities feel when they are acting as representatives of their perspective groups. My friend said that sometimes she needed to be a good actress, and I agree with her. When I didn't see anyone that looked like me around the room, it preyed on my fears and made me ask myself, "Should I be here?" I felt like I was being challenged to prove that I belonged, so I pretended and acted competent. I didn't have a lot of self-confidence for years: I had to act and pretend because there was no other alternative. It takes a lot of energy being someone who you are not, and you constantly worry if someone is going to figure you out. These issues of self-confidence affect men, too, but are amplified for women. Some people are naturally self-confident, but [for] a woman in a scientific workplace, it helps to act that way. Eventually I became more confident, but I want to make it easier for underrepresented groups in STEM fields—make them feel welcomed from the beginning.

During my high school and college years, I had the opportunity to participate in various geology internships. I completed an internship at the US Geological Survey helping out in the lab, which I loved. While in college I interned at the National Air & Space Museum and was assigned to use Landsat images to map areas on Earth. I also completed an internship at the NASA Jet Propulsion Laboratory (JPL), where I mapped an area on Mars. These experiences were fun and convinced me I was on the right career path.

I was fortunate enough to have NASA associate administrator and geologist Tim Mutch as a mentor while I was in high school. Tim worked with my father at NASA headquarters for several years. He was amazing! Tim gave me great college and career advice that I still use today: he recommended that I go to a university with a traditional geology department and study the geology of Earth if I wanted to be a planetary geologist. Tim had been a geology professor at Brown University, where he wrote books on the geology of the moon and Mars. He also was an explorer on Earth and unfortunately died while on a climbing expedition in the Himalayas in 1980.

I listened to Tim's advice and attended the College of William and Mary in Williamsburg, Virginia, and majored in geology. During my matriculation,

I had two important mentors, paleontology professor Gerald Johnson and petrology professor Steve Clement, who was my research adviser. My senior thesis focused on how one area in Virginia had changed over time due to cycles of heat and pressure. I examined the minerals in the rocks to understand the conditions and sequence under which they had been formed. It was interesting and fun lab work, requiring lots of time looking through a microscope. I was able to determine what environmental conditions that area of Virginia had been subjected to.

The opportunities and experiences I had while pursing my undergraduate degree in terrestrial geology and the time I spent in the field have certainly helped me in my career as a planetary geologist. When I look at a planetary image and draw conclusions about how a planet evolved, I know that we can crawl around an area for 100 years with rock hammers and still not get it quite right—that thought keeps me humble.

GRADUATE SCHOOL AT BROWN STUDYING VENUS

I knew early on that I wanted to get a PhD in geology. I visited several colleges when I began my search for graduate school, and I selected Brown University in Providence, Rhode Island. I chose Brown in part for sentimental reasons because my mentor, Tim Mutch, had been a professor there and had died tragically a few years earlier. However, I was also impressed with the environment—I liked New England, and I had a good feeling about the midsize Geology Department, where all the professors seemed very engaged.

When I started at Brown, there were 20–30 geology graduate students, nearly split between women and men. By the time I left, more than half of the women were gone. Some left because they weren't sufficiently interested in getting a PhD. Others realized that graduate school was also a tough and competitive environment. Many of the women who remained in the program and were successful had been athletes in high school or college and were very competitive. I am not very competitive, so I struggled.

I was lucky, though, to have an incredibly supportive husband while I was in graduate school. I had met him at William and Mary, and we got married in 1984, a year after we graduated. While at Brown, I came home several times and shared with my husband the intense pressure I was under, with criticism sometimes seeming to come from all sides. My husband asked me if the critics were right, and I replied, "They are completely wrong." Then he gave me advice on how to approach the situation with the critics. Unfortunately, many of the other women who ended up leaving the geology graduate program did not have someone like my husband in their lives. Throughout my career, my husband has been 100 percent supportive of what I do—not propping me up but more accurately not letting circumstances derail me.

> My advice for students who are making research choices is to ask [themselves], "Where [is] a field ... going, and if I want to be at the forefront of it, where should I position myself?" For example, if I were making a choice today, I would study the surface of Mars or search for extrasolar potentially habitable planets—those are the hot fields going forward.

For my thesis, I studied a type of feature on Venus. When I got to graduate school, the Soviets had a satellite mapping the northern half of the planet, which was the first time we had high-resolution data of Venus's surface. Brown University and my adviser, James Head, had a close relationship with scientists at the Vernadsky Institute in Moscow. Data from the satellite showed enigmatic corona features that were similar to volcanic hot spots in Hawaii but that had never been seen before on Venus. I wrote my dissertation describing the theory for how those features formed.

The opportunity to work with these data and to write fundamental papers on the geology of Venus gave me exposure. I presented at major conferences describing these features, which are very prominent on the surface. Whenever there was a request for a woman speaker to talk about Venus, I was often selected. I am sure they liked my work, but I was also in the right place at the right time.

When it was time for me to graduate from Brown, I started inquiring about a full-time job. There initially didn't seem to be any suitable positions available. I inquired at Goddard Space Flight Center in Greenbelt, Maryland, and at JPL since I had completed internships at both centers. During that time, the United States had a spacecraft called *Magellan*, which was a mission to Venus. When I wrote to the JPL project scientist, whom I had met a number of times, and asked if there were any open postdoc positions, I was offered a National Research Council postdoc there to be part of the *Magellan* team. At the time, my husband was working at a bank in Philadelphia, but he agreed that it was a great opportunity, so he quit his job, and we moved to California. He found a new job that turned out to be much better than his previous job. Approximately four months after I started working at JPL, the *Magellan* probe went into orbit around Venus.

SEXISM AND OLD ATTITUDES

During graduate school, I was conscious that there were very few women in my professional field, which made me feel as if I had to work twice as hard to be taken half as seriously as the men. My graduate school adviser gave me opportunities to speak at team meetings so that my voice would be heard. While I was working on the *Magellan* mission at JPL, I presented updates and participated in discussions, and sometimes I felt that people seemed impressed that this "girl" could string two sentences together. They weren't rude, but they were condescending, and I felt as if they didn't take me seriously because I was a woman. I felt as if I had to outperform everyone else, which is not a great way to feel. However, I never had a professional situation during grad school in which someone was openly discriminatory.

Before I was offered the job at JPL, I told a male classmate that I was hoping for a position, and he replied, "It doesn't matter if you get the job or not because you have a husband." I was extremely angry, and I thought to myself that the guy was biased. Fortunately, that interaction happened a long time ago, and attitudes have changed some since then.

My first three years at JPL were amazing. Every day we mapped areas of the surface of Venus for the first time at high resolution. Again, I was lucky to be in the right place at the right time. One of the ways to be a successful experimental scientist is to pick a field where there is a great deal of new data to analyze. When I moved to California, I thought maybe I should have chosen a career in seismology instead because at the time there were a lot of earthquakes in southern California, so an endless amount of data was being generated! However, there's more to being a successful scientist than just having access to data. As a scientist, you need to constantly question your ideas, as well as everyone else's.

I worked at JPL for ten years, from 1990 through 2000. I started on the *Magellan* mission as a postdoc and later became the deputy project scientist. I then worked on a radar-imaging instrument for the Space Shuttle that flew twice in 1994, a project that included training two astronaut crews. I also worked as lead scientist on the New Millennium Program, which was established to develop and test new space-exploration technologies. By 1995, this work became mostly program management and less science, which I really love and missed, so I decided to take a break from management and spend more time with my young kids rather than flying to meetings across the country. I reduced my work time to 30 hours a week, and I became a college professor. However, I had to leave the United States to make those changes.

RELOCATING AND REBALANCING

My family and I relocated to London for five years because of my husband's job assignment. While I was there, I taught at University College London, advised graduate students, and conducted research. I remained employed by JPL and worked on NASA missions but did much less management and more science.

At JPL before we made this relocation, I had tried reducing my hours, but it was difficult. Although the management at JPL was supportive of my desire to work part-time, I found it hard to discipline myself to stay home. When I was asked to work on interesting projects, I had a hard time saying no; and when my colleagues asked me to come to meetings on my days off, I didn't have the courage to tell them I was off. So when my husband was informed of the opportunity to go to London, I was relieved and ready to go.

While in London, even though I was working fewer hours, I was able to stay disciplined and remain connected to the profession. I tried to stay visible by keeping my publication rate up and attending conferences. My husband did a lot of traveling, but we were able to divide our family responsibilities so that I could travel to advisory committee meetings or attend review panels. Even though I had stepped back from some of my professional responsibilities, I hadn't stepped out. Some of my colleagues didn't realize I was still working as a scientist. Every once in a while I would attend a conference, and a colleague would say, "Oh, I thought you quit the field."

As my kids got older, they required much less time and energy, so I returned to work full-time. In 2000, I was asked to be an associate member of the Cassini mission to Saturn radar team. In 2003, we started receiving data from Titan, and I fell in love. I know it seems strange to fall in love with one of Saturn's moons, but I did. Titan is approximately 750 million miles from Earth, and it rains methane at extremely cold temperatures. As a geologist, I was amazed to find an Earth-like world at the other end of the solar system. In 2005, I helped to formulate the arguments that the radar features at Titan's north pole had to be lakes and seas; these arguments became the outline of a *Nature* paper announcing the discovery, of which I was first author.

Shortly after the *Nature* paper was published, I got a phone call from a colleague at Lockheed Martin. They were considering submitting a proposal to send a floating probe, a boat, to one of the seas on Titan, and they wanted me to be principal investigator. I thought the likelihood of this $425 million mission being funded was low, but I was in love with Titan, so I agreed. We were able to prove that Titan could be explored at relatively low cost. I worked on the proposal for five years, and the mission was close to being accepted. We made it through multiple rounds but lost in the final round of NASA's Discovery Program.

My team and I put in a great deal of work on the Titan proposal, so losing it was a huge disappointment. When I was notified that it wasn't selected, I wasn't sure what I would do next—perhaps finish up some Venus and Titan papers and look for a job in science or policy. Shortly after this disappointing news, I received a phone call asking if I wanted to interview to be the chief scientist of NASA. I was appointed to the position in 2013. I never desired to be the chief scientist because I really wanted to lead the Titan mission; however, the job is a wonderful opportunity, and I am really enjoying advising and leading in this role. I have the amazing opportunity to be involved in all of the science NASA does and to reach out around the world, talking to students about the challenge and thrill of exploration.

ABOUT THE SUBJECT

Ellen Stofan is a planetary geologist and was Chief Scientist at NASA from 2013–2016. Since the printing of this book, she is Director of the National Air and Science Museum in Washington, DC. Stofan holds an undergraduate degree from the College of William and Mary in Williamsburg, Virginia, and master's and PhD degrees from Brown University, all in geology.

33 BREAKING BARRIERS AND LEADING TEAMS

Ellen Pawlikowski

No sooner had the words left the one-star general's mouth than I said to myself, "I'd *love* to prove her wrong."

The year was 1983, and I was a first lieutenant at McClellan Air Force Base, near Sacramento, attending a mentoring meeting in which the visiting general officer would share her unique career advice with other impressionable women officers. That day, a grand total of three of us were in attendance—that's right, there were only three women officers on the entire base.

The general spoke to her career challenges, saying that for us to achieve success like hers, our lives would undoubtedly include sacrifice: marriage would be difficult and children an impossibility. I left the meeting angry, upset, and full of resolve. My male counterparts, doing the exact same military job as I wanted to do, could have a family. So why couldn't *I* achieve career and personal success, too?

Twenty-two years later, in a ceremony pinning on my new rank as brigadier general and with my two daughters by my side, I asked that one attendee stand for recognition. In the audience that day was that same one-star general of years past, now long retired, who had become a role model of mine. That day in June 2005, I presented her with a dozen roses, gave her a hug, and proudly said, "I've been waiting 20 years to prove you wrong."

As an air force leader now charged with overseeing 80,000 personnel at installations across the country, I feel it's important to share with others that seemingly insurmountable challenges can also transform into the

Ellen Pawlikowski (*far right*) and her sisters, Mary, Joan, and Kathy, grew up in Newark, New Jersey, in the 1960s.

greatest of opportunities. My path, like that of any other, has certainly been strewn with obstacles, but the courage to commit to myself and my goals has seen me through time and again.

STUDENT JOURNALIST TO TEAM SCOREKEEPER

I grew up near Newark, New Jersey, the second of four daughters. My father was a first-generation American from a Polish immigrant family—his parents had come from Poland to Scranton, Pennsylvania, where my grandfather had worked in the coal mines. My dad served in the Japanese Occupation Force after World War II, then went to school on the GI Bill to become a high school social studies teacher. My mother made her living as a secretary, first in industry and then at Rutgers University until her retirement.

In high school, I wanted to be a journalist—I was the editor of the student newspaper and the assistant editor of the yearbook. When my dad realized I was leaning toward journalism as a career, I remember his telling me, "You need a degree you can use to support yourself. You won't be able to survive as a journalist." He steered me instead toward classes that would make me use different skills. In my sophomore year, while all the other girls were taking home economics, I took a drafting class. Though he encouraged all of us to follow that technical path, I would be the only one of my sisters to become an engineer. In many ways, I became the son that my father didn't have.

The drafting, chemistry, and physics classes that I took in high school were not very exciting, and I learned the material by rote. But I loved math, and my sophomore-year geometry teacher noticed my interest. Realizing that I was outperforming my peers, my teacher sent me home that summer with an Algebra I textbook. I taught myself algebra that summer so that I could join the accelerated track the following school year. My junior-year math teacher also coached the junior-varsity baseball team, and he asked me to be a scorekeeper and statistician for the team. When I told him I didn't know the first thing about baseball, he replied, "Well, you are going to learn."

I spent that spring sitting next to him on the bench as he taught me the theory of baseball. Should we shift the infield one way or another for a particular hitter? What should be our pitching strategy? It was my first exposure to applying math to solve what was essentially an optimization problem.

COMMUTING TO COLLEGE

My sisters and I understood early on our parents' expectation that we earn college degrees and forge careers to support ourselves. We also understood that our parents didn't have the money to send us away to college, so we could live at home for free, but we would have to earn our own money for tuition, books, and transportation. When it came time to apply for college, I looked for schools within commuting distance and decided on the Newark College of Engineering (now the New Jersey Institute of Technology). Even then it had the reputation of being a good "blue-collar" engineering school— the kind whose engineers could walk out the door, get a job in the chemical-processing industry, and have an immediate impact. I also chose Newark College because it was close to Rutgers, and I still hoped to continue my writing education there and maybe earn a journalism degree simultaneously.

The Vietnam War was in its twilight years in 1974 when I started college, and I had seen a lot of press about the military—some of it not so good. I joined the Reserve Officers' Training Corps (ROTC) during my first year of college, mainly because I was curious about the military. At the time, women in the air force couldn't be pilots or navigators and therefore weren't eligible for ROTC scholarships. It became pretty commonplace for me to note that in ROTC at a male-dominated engineering college, I was often the only woman in the room. I would be the first woman from my detachment to graduate and be commissioned into the air force. In the

ROTC, though, I also found that I enjoyed the camaraderie and the sense of being part of something larger than myself. At a commuter school, without a dorm to go back to, the ROTC lounge became my home away from home.

ROTC also became an extended family for me—quite literally. The man I would eventually marry was two years ahead of me in the detachment, and another member of the detachment married my husband's sister. All of us in the detachment have kept in touch through the years, even as many of us left military service. A whole contingent from my ROTC detachment drove down to join me at my three-star and four-star promotion ceremonies—they were likely the rowdiest group there.

> When I was growing up, the expectation was that many of us would stay [to work] at the same company for decades. This is less true today—young people tend to try something for a few years after graduation and then do something completely different. But I found a similar level of flexibility within the air force—from my first posting to now—with every reassignment, I was learning the ropes on a new project with a new set of challenges.

I worked in retail and as a waitress to earn tuition and spending money, and the experiences taught me that neither was a good career option for me. After working during the summer and then part-time for a private chemical company, I found I was happier doing chemical engineering. When that summer job ended, I remember asking myself, "Do I want to use my talents to help a company make money, or do I want to be part of a cause connected to national security and bringing home people safely?" That was when I decided I would get my commission and commit to four years of military service. I didn't necessarily think I was going to be a career officer, but it would at least give my career as an engineer a start. It helped that as the Vietnam War wound down, the air force no longer had such a pressing need for pilots. As a result, it started awarding scholarships to other career fields, and, after signing an ROTC contract my sophomore year, I was awarded a scholarship the following year.

During my junior year, I started to develop a love for thermodynamics. Plenty of people—even chemical engineers—groan when they think about their first introduction to thermodynamics, but for me it was a perfect example of using theoretical mathematics to solve real-world problems. One of my professors introduced me to a colleague, David Sukovich, who worked at a chemical plant. Sukovich was solving problems of chemical equilibrium in distillation columns, steam strippers, and other separation processes. I worked with him at the plant's corporate office, solving such problems using thermodynamic methods.

I hadn't planned to continue to graduate school after college, but when I returned to Newark College for my senior year, Sukovich and some of my professors challenged me to go. It was their opinion that I was a good enough student that I wouldn't have to pay tuition to attend. I asked the air force permission to defer service, and, with its continued downsizing, it agreed to let me attend graduate school—at least for a master's degree.

In the 1970s, the newest area in thermodynamics was called "molecular thermodynamics," which combined the classic theory of continuous gasses with modern information about the behavior and conformation of individual reacting molecules. I visited one of these molecular thermodynamicists, John Prausnitz, at the University of California–Berkeley and fell in love with the research area. Berkeley wanted me to apply as a PhD candidate, so I requested another two years' deferral from the air force, which it granted, but another difficulty soon presented itself.

Paul and I had been dating since my sophomore year, and we were a serious couple. After he graduated, he attended navigator training and eventually took an assignment at McClellan Air Force Base near Sacramento, California.

Ellen Pawlikowski (*second from left*) and members of her ROTC detachment at the Newark College of Engineering (now New Jersey Institute of Technology) were commissioned into the US Air Force in 1978.

So I was glad to be moving to Berkeley for graduate school—it was much closer to Paul than Newark. However, Paul and his mother especially were dismayed by my decision to attend graduate school. They were traditional in the sense that they expected me to finish college and then stay at home and travel with Paul to his air force assignments while he supported me. We had to do a fair amount of soul searching to keep our relationship going.

THERMODYNAMIC RESEARCH

Most students in the Berkeley PhD program are admitted without any guarantee of who they will work for, but I had been so compelled by Prausnitz's work that I asked for and received upon my admission unofficial assurance that I would be working for him. Having this certainty helped me quickly start my research.

The end of the 1970s brought the last great energy crisis, and one of the more intriguing ways out of the crisis was coal gasification—turning coal into natural gas to replace other expensive hydrocarbons. But the coal we were gasifying wasn't pure carbon, and gasification created poisonous contaminants such as hydrogen sulfide, ammonia, carbon dioxide, and hydrogen cyanide. The contaminants had to be removed from wastewater before it could be recycled or dumped into a river. Because these contaminants are weak electrolytes, meaning that they exist partially as ions in water, it is difficult to remove them from the waste stream.

For my thesis, I used principles from molecular thermodynamics to understand the energetics of removing weak electrolyte contaminants from wastewater. The experimental portion of the work made good use of my "blue-collar" engineering degree. We created a model based on an understanding of molecular structure, which we parametrized based on experiments with contaminants at high temperature and pressure.

One of the early difficulties of my marriage was overcoming the expectation that I wouldn't work at all so I could stay home and take care of the kids. In fact, when my husband left the air force and started teaching, he ended up taking responsibility for carpooling and childcare in a way that allowed me to keep moving my career forward. He was brought up thinking that the wife would stay home and take care of the kids, but we were able to flip that script almost completely and make it work. So one of the things I learned over three decades of maintaining a marriage and a mobile career was to make sure we kept an open mind and were willing to accept change.

Paul proposed to me at Thanksgiving during my first fall at Berkeley, but we agreed to put off getting married until after my qualifying exams. In the 1980s, the air force didn't yet have a formal spouse system, and Paul was about to be transferred. In order to be stationed near Berkeley, Paul promised the noncommissioned officer picking his next assignment a dozen roses if she would keep him in Sacramento. The approach was successful, and he got to stay in Sacramento. After I completed my thesis at Berkeley, I was also assigned to McClellan Air Force Base in the gas research and development group of the Technical Operations Division.

My first assignment was in the lab responsible for monitoring the Nuclear Test Ban Treaty. Part of our work required analyzing air samples from the Sea of Japan and elsewhere, looking for krypton and xenon isotopes, which would indicate a recent nuclear test. My first task was to design an apparatus to separate those gases from air and then to measure the isotopes.

I had a drafting table and drew out a plan for all the equipment needed to build this separation apparatus. A team of people would be working under me to make that happen. That was my first chance to explore what I could get done if I were leading a group rather than doing everything myself. The ability to sit down, think about and develop an approach, and then empower my team to carry out the details was very rewarding. Not

Ellen Pawlikowski (*third from left*) and members of John Prausnitz's research group at the University of California—Berkeley head out to lunch, circa 1980.

only did I enjoy leading a team, but it also turned out I was pretty good at it. So that's how I migrated to the management side rather than continuing as an individual contributor.

A LONG AIR FORCE CAREER

I spent most of four years at McClellan in the nuclear-test-monitoring lab, a period that would just about satisfy my commitment to the air force. Thanks to my work in thermodynamics and energy processing in graduate school, I continued to get calls from companies wanting me to join them after I finished my air force assignment. But I really liked my job and felt that I had the resources to do anything. Rather than leave the air force, I stayed in for another four years, then another four, before eventually deciding that I would be a career officer.

Because both Paul and I changed air force assignments regularly, we couldn't always depend on floral gifts to assignment officers to keep us together. After my second assignment, Paul had to go back to flying B-52s, and I was promoted to major. At that time, I was selected to attend Air Command and Staff College, and I took our two daughters with me to Maxwell

Ellen Pawlikowski's husband, Paul, arranged for both of them to work at McClellan Air Force Base in California after they were married in 1982.

I met a few mentors early in my career who inspired and advised me. Ed Giesselman, the colonel in charge at my first posting in the [nuclear-test-monitoring] lab, was one of the first to help me understand the many opportunities the air force would give me to lead and explore new science. He is a big reason I took my second assignment. Paul Nielsen, whom I worked with at Rome Lab [in New York State], served as a model for me in how to lead scientific teams in the military and is still someone I consider a mentor and friend nearly 30 years later.

Air Force Base in Montgomery, Alabama. For a year and a half, I worked and studied full-time while raising two children. That experience has given me a huge appreciation for single parents.

While we were at Rome Lab after Desert Storm in 1991, the air force started downsizing again. They offered Paul an early out, and we realized that it would be much easier to stay together and keep focusing on our relationship if one of us left the air force. Paul had always wanted to be a teacher, so we made the decision that he would leave, and I would stay in for my 20 years. Paul became a math teacher and took on more of the responsibility for raising the children. Without that decision, I would not have been able to do what I have done.

I'm sometimes asked if I regret not using my PhD, as if the only use of an engineering PhD is to work as an engineer in precisely the same field. My response is that I use it every day! The most important skill I [could] learn from [my] PhD is the ability to enter a discipline I know nothing about and learn enough about it to make a difference. Without that ability, I wouldn't be able to effectively lead technical teams, nor would I have been as effective a bridge between technical and operational teams.

The rest of my career has been characterized by moving to a new place, starting a new assignment, and working to bring together a team to make the science happen. To be an effective leader, I learned it was less important that I initially understand all the technical and scientific points and more important to build rapport so that people wanted to follow me. It took me some time to fully internalize that what motivates scientists and engineers—whether civilian or not—is not always the same as what motivates others—for example, a pilot flying combat missions. The higher I

Ellen Pawlikowski and her team at McClellan Air Force Base analyzed air samples by mass spectroscopy to monitor the Nuclear Test Ban Treaty, circa 1984.

climbed the air force ladder, the less I saw my job as getting the mission done directly and the more I saw it as making sure my team members had what they needed to complete their mission. I didn't learn that in engineering school or working in industry—I learned leadership through observation, experience, and mistakes as I moved from one assignment to another.

Another leadership role that I have learned during my air force career is how to act as bridge between technical and operational personnel. I can sit down with scientists and engineers and later meet with my fellow four-star generals to translate highly technical information into terms they can understand. This is a talent not readily available among personnel in the air force, and it was very useful when I led the Airborne Laser program from 2000 to 2005.

In this program, we were designing, building, and installing a megawatt laser and sophisticated optics onto a Boeing 747, with the goal of shooting down ballistic missiles during boost phase. Whereas the scientists understood and described the laser system by technical parameters such as efficiency, thermal load, and jitter, the weapon operators were concerned about how to aim the system and its maximum range. Because of my scientific background, I could translate between the two groups. I could also act as a bridge to my bosses, allowing them to understand the program's challenges and progress.

It's been quite a few years since then. When I got my historic promotion to four-star general, it was almost surreal being interviewed by the press. In high school and college, I had dreamed of being a journalist interviewing important people, and instead now I was the interview subject! Life is so full of opportunities, despite what can initially seem challenges, and I know I couldn't have ever predicted this is where I would end up.

ABOUT THE SUBJECT

General Ellen M. Pawlikowski retired from the US Air Force in 2018. She most recently served as commander, Air Force Materiel Command, Wright-Patterson Air Force Base, Ohio. She is the third woman to achieve the rank of four-star general in the US Air Force's history. General Pawlikowski has a BS in chemical engineering from New Jersey Institute of Technology and a PhD in chemical engineering from the University of California–Berkeley, where she was a Hertz Fellow.

34 ENGINEERING A LEADER

Paul Nielsen

I had no idea what I was getting into when I arrived, fresh from getting my PhD, at the US Air Force Office of Special Projects in 1981. The SPO (as we referred to it) was actually the National Reconnaissance Office, and at the time everything about the office, even its existence and the acronym NRO, were classified. My superiors had recommended the SPO to me as an exceptional posting, but when I asked what my job there would entail, all they would say was that I would find out when I got there.

When I arrived, the lieutenant colonel who would be overseeing my work told me he didn't think they needed a person like me, a PhD scientist, liable to spend all my time thinking and analyzing. Instead, he said, their group needed engineers to build and deploy actual hardware. I soon found out that my supervisor, who also had a PhD, was teasing and perhaps testing me. I answered him by saying that I didn't expect everything we would build to be perfect on the first try and was looking forward to working with the NRO team to reengineer the hardware as we went along. That was a good answer at the time, and I subsequently found that my PhD in physics was excellent preparation for engineering at that first job and throughout my career.

A CHALLENGE AND A GIFT

In the third grade, I had a teacher who made a tremendous difference in my life. I grew up in New Orleans, Louisiana, until that year, when my family moved to a recently opened suburb called Covington on the other side of Lake Pontchartrain. About six weeks after moving, I was still feeling out of place when our English teacher told the class about a reading contest. I was a pretty good reader and thought this would be my chance to shine, that people would notice me and maybe I could make some new friends.

At the end of the day, our teacher took me aside and told me that I wasn't going to get to compete in the reading contest. I felt crestfallen. But then she said that she had a special book for me to read. It was a long book and so would take a long time to read, which was why I wouldn't be able to participate in the contest. Then she gave me an unabridged edition of Homer's *Iliad*, thankfully in an English translation.

> In the purest delineation of their roles, scientists search for new knowledge, and engineers build things and solve problems. That said, I think that studying a pure science like physics gives you the background to be a great engineer. It helped made me think about the system as a whole. Just like conservation laws in physics, I learned to understand about trade-offs—if you do something here, it's going to cost something there.

I was a pretty determined kid, so even though it did take a long time, I made my way through the *Iliad*. And I understood a fair amount, though probably not as much as if I had read the epic poem as an adult. This experience was a gift; I learned that if I could read and understand the *Iliad*, I could read anything. I later spent many of my summers in the library, reading widely whatever seemed interesting.

I always knew I liked science and math; even as a kid, I liked to take things apart and try to put them back together—not always successfully. But I didn't know what it might be like to become an engineer or a scientist or a mathematician. I just didn't know anyone who did that. My sophomore year, a cadet at the US Air Force Academy who had graduated from my high school came back to the school to talk about what it was like to study science there. I got excited because I could imagine myself in his shoes. I ended up applying to only two schools—the Air Force Academy and MIT. Even at that time MIT was expensive, and finances mattered to our family, so when I was accepted at both institutions, I chose the academy and headed to Colorado Springs.

My first year at the academy was very challenging; the combination of intensive technical courses with military training almost felt like hazing. I did well, even though I couldn't decide on a major. I knew I wanted it to include physics but was never happy with just a single topic and switched back and forth between joint majors of physics with math, engineering and computer science.

The summer after my junior year, I had the opportunity to do two months of research at the Lawrence Livermore National Laboratory. While there, I got a call saying that Edward Teller wanted to talk to me and that

Paul Nielsen (*right*) attended the change-of-command ceremony of the Air Force Research Laboratory from Ellen Pawlikowski to Neil McCasland in 2011. Nielsen was the lab leader prior to Pawlikowski, from 2000 to 2004.

I should go to his office. I knew Teller was the famous physicist popularly known as the "father of the hydrogen bomb" and felt intimidated. When I got to his office, he told me about the Hertz Foundation and the Hertz scholarship and asked if I was going to apply. I said I was planning to apply for a Hertz and for a Rhodes as well. He said, "Ah, if you get both, you must turn down the Rhodes. They don't know science in England!"

I was awarded a Hertz Fellowship, so after graduating from the academy, I returned to California to get a master's degree at the University of California (UC)–Davis, while continuing my research at Livermore. After I finished the master's, the university wanted me to continue and get a PhD, but the Vietnam War was heating up, and the air force wanted me to give something back for all the years of free education they'd funded, so I spent five years working for the air force.

MENTORSHIP AND EXPERIENCE

It was a challenge taking that five-year break from my PhD. When I returned to UC Davis, I had forgotten many things that used to be second nature—for example, I had to look up integrals that I used to know by heart. By that time, I had also married and had a son. And after having a regular job, I was not used to slavishly having to think of my projects and my thesis every waking moment. But I did some pretty interesting things in those five years.

My first assignment was to the National Security Agency (NSA) in Maryland working in communication security. The NSA called me a reliability physicist, and the job was fascinating. I worked on qualifying parts that contractors were building to put in NSA satellites. I would visit the vendors that were making early integrated circuits and other devices, and I remember marveling when a vendor had developed an integrated circuit holding 256 bits. During that time, I visited contractors such as Harris Corporation and Fairchild, some of whose employees went on to found Intel.

Later, when I was working at the air force's weapons lab at Kirtland Air Force Base in New Mexico and was almost ready to return to my PhD, my commander called me into his office. He told me that General Lew Allen, who would later be the head of the US Air Force, had just become a four-star general and was looking for an aide. I figured I would apply since I wasn't going to get the job anyway, but I did get it, which added another year before I returned to school.

I wouldn't have worked well with most generals, but Lew Allen was a great fit for me. He had a PhD in physics, had been head of the NSA when I was there, and was a member of the National Academy of Engineering and a fellow in the American Institute of Aeronautics and Astronautics. So he was a great role model for me, showing that a scientist or an engineer could get ahead in the air force. He also taught me a fair amount about leadership.

I was Allen's aide for only nine months, yet during that short time I met many high-ranking officers. I learned that generals are regular folk, not really different from the rest of us. They come from different backgrounds and succeed by working hard and keeping their nose to the grindstone. I watched General Allen lead not by barking orders from behind his desk but by discussing solutions and acknowledging when he was wrong. I kept in touch with many of the people I met during my time with General Allen, which turned out to be important in advancing my career. One of the officers I met was the colonel who recommended me for the SPO assignment after I finished my PhD.

My PhD thesis was in the area of computational plasma physics. I ran simulations on a CRAY supercomputer, debugging massive amounts of undocumented code written by academics to model magnetic plasma confinement. At the time, we thought it a major advance that we could

Paul Nielsen was promoted to two-star general in the US Air Force in 2001.

improve on one-dimensional models by running two-dimensional simulations. Now, of course, computers simulate three-dimensional physics models with incredible detail, often in less time than the early one-dimensional CRAY simulations required.

An interesting difference between working for the air force and working in an academic job is that you can do a PhD in one area and then be assigned a job in a completely different area. So after completing my thesis in plasma physics, I went to SPO to design spacecraft.

Working at SPO was a wonderful experience; it was a place full of hand-selected people doing advanced work and collaborating with the best of the air force contractors to build wonderful spacecraft that really helped our nation. First, I worked on satellite "bus" technologies, the craft's power source and positioning systems. Later I became a chief engineer for the satellite constellation. That job was probably the high point of my technical career because after I left this project, I started to be assigned management jobs.

MANAGING TECHNICAL TEAMS

It's hard to remain too technical in the air force. The military needs high technology but depends mainly on the contractor base to provide it. Once you get to be more senior, you need to move into management in order to advance. I was already inclined in that direction—I had known since I played team sports in grade school that I liked working with people. I had stopped playing team sports in college after I met my roommate—a guy my height, but 60 pounds heavier who wrestled and played running back for the US Air Force Academy—but I still knew I liked being on teams and working with people.

I also have strong interests in nontechnical areas outside of science. I love literature, writing, and history. Having both technical and people skills helped me stand out in the air force and get promoted. Sometimes the air force thinks a good manager or even a fighter pilot can manage everything, but to manage a technical team and make good decisions, you need to have a good intuition for technical trade-offs, and I had that.

It is common in the military to move around, and I moved a lot in the decade after SPO. I worked on missile defense, satellite tracking, and many other small projects. I eventually ended up at the Pentagon, managing satellite communications and surveillance. Most of that work was on the Milstar constellation of satellites, which started launching in 1994 to

When it was announced that I would lead the Rome Air Development Center in Rome, New York, a general at the Pentagon named James Abrahamson took me aside and asked if I'd ever managed anyone. I said, "Sure, I've managed some offices with 40 to 60 people." He said, "Managing this large an organization is going to be different." He was right, and many managers have trouble making that transition. You have to learn how to manage people through other people. One of your key jobs is to pick good managers and to support and cultivate them. It's good if you get to know them very well, as I did in Rome. I learned to motivate them, and it was great to have so many sharp people working for me.

provide secure and reliable communication for the air force around the world.

By then I was a colonel, with two decades of service behind me and thinking of retiring from the air force. My kids were in high school, and DC had many good jobs for my wife and me. I had been at the Pentagon so long I wasn't even sure the air force still knew I existed. But then they found me again.

The air force offered me a tempting job in an interesting place: command of the Rome Air Defense Laboratory in Rome, New York. I would manage about 1,000 people and a budget of $400 million per year, far more people and money than I had ever managed before. In Rome, I learned more about leadership and about how to delegate responsibilities. I learned how nice it is to have independence—my boss was hundreds of miles away in Boston.

A DIFFERENT SORT OF COMMAND

In December 1994, I was summoned to General Roland Yates's office. When you're summoned to meet a distinguished commander, you drop everything, get on a plane, and go—one day I was working in New York, the next day I found myself in Ohio. I recall his office was intimidating and full of memorabilia and the man powerfully charismatic. He told me that he and his friend Joe Ashy—another general, then in charge of Space Command—had talked and decided I needed an operational tour. They wanted to send me to Cheyenne Mountain in Colorado to be one of the shift commanders there.

The idea of leading an operational team scared me—I would be leading the actual day-to-day capabilities of the air force rather than just overseeing research. But the actual job turned out to be pretty technical; since I had helped develop many of the air force's satellite and communications systems, I knew how such operational teams were supposed to work. And having a diversity of command experiences is important in the air force: it was in Colorado that I was promoted to general.

As a general, I was eventually assigned to Kirtland Air Force Base in Ohio, where in 2000 I became the commander of the Air Force Research Labs (AFRL). It felt as if I had been groomed for that job all my life. Many of the previous labs in which I had worked—in Rome, at Kirtland, and others—were all under my command now. We had people at 50 different places around the world. I ended up staying at Kirtland for four years, which is a pretty long command in the air force.

STAYING YOUNG

In the air force, they say you know when it's your time to retire. After four years at AFRL, I knew it was my time. I didn't think there would be another job in the air force more fun or better suited to my background than leading AFRL.

When the job to direct Carnegie Mellon University's Software Engineering Institute opened up, I thought about my mentor General Lew Allen. When he retired in 1982, some people had thought it strange that he went from being chief of staff for the air force to being director of a small lab at JPL. But it was interesting work. So I thought of that when I retired; leading a smaller organization seemed like the right choice.

At the Software Engineering Institute, we support the military and parts of the rest of the federal government to build software and help keep it secure. In addition to designing secure software, we also look at issues of security in the supply chain by reverse-engineering malware and by doing cyberforensics. In the latter area, we also help local law enforcement office when they need to investigate cybercrime.

Working in the air force kept me young because when I was a 50-year-old, most of the people I worked with were still 20-year-olds. My new job at a university feels the same way. I like that; it keeps my enthusiasm and curiosity high, and I hope I can keep at it for a long time.

ABOUT THE SUBJECT

Paul Nielsen is director and CEO of the Software Engineering Institute at Carnegie Mellon University. He previously served in the US Air Force for 32 years in a variety of roles, including commander of the Air Force Research Laboratory and as the air force's technology executive officer. Nielsen holds a BS degree in physics and mathematics from the US Air Force Academy and both an MS in applied science and a PhD in plasma physics from the University of California–Davis, where he was a Hertz Fellow.

35 VISION AND TEAMWORK

John Mather

Over the course of my career, there have been many times when experiments didn't work, and I wondered what I should do next. One of those times was after my graduate thesis experiment failed. We built an instrument to measure the spectrum of the cosmic background radiation and sent our instrument high above the atmosphere on a balloon. Unfortunately, we were rookies in the balloon business and had not tested everything we could possibly test.

After that failure, I thought it was obvious that the experiment was too hard for me, and I should really try something else. I went to the NASA Goddard Institute for Space Studies in New York to learn to be a radio astronomer. That, too, is a hard task. But when the opportunity arose to propose an experiment, I said, "Let's try my thesis experiment again, only this time in space." The experiment was sufficiently difficult that no one else was going to try it, and it eventually led to the Cosmic Background Explorer (COBE) satellite, which helped cement the Big Bang theory of the universe.

The reason the COBE succeeded was teamwork. When we proposed COBE, I knew that if we were chosen, NASA would help us get the proper engineering behind it. For the right team, the experiment wasn't too hard. You get the help you need.

Teams succeed with the right combination of individuals who possess a full complement of skills, including personnel management. This is described in a book I highly recommend—*How NASA Builds Teams* by

Charles Pallerin. In a NASA team, we need at least four different personality types. I am cited in the book as an example of the "visionary type," which is someone who thinks far into the future, imagining what might be done, and how a team might get there.

The visionary doesn't necessarily have to be good at implementation, but the team needs people who are. They figure out what should be done first and what should be done second. They estimate how much work the experiment will take and how much it will cost. They are the people with checklists. You also need team members who will manage using people skills to make sure that the team members are at least half-happy with each other.

When Charlie Pallerin was writing his book, I told him that half the price of a mission is socially determined. You can easily double the cost of a project if the people process doesn't work right. For example, you can double the cost if the people making the decisions don't know what to do or decide incorrectly or are unable to attract the right kind of talent to solve problems. The doubling factor I gave to Charlie was a made-up number, but Charlie took it seriously, and I think it reasonably true.

I know it to be true because I have spent time as a manager. At NASA, management means choosing people to work on a project, writing their performance appraisals, getting them promotions, and on rare occasions firing them. Being a manager means figuring out which office they get to sit in. There are many things that seem to have nothing to do with science but are really important to making an organization or a project function. I have done some of these things and have enjoyed the process. But I have also found management difficult, and it isn't really where my heart is.

The good thing about being a manager is that you get to work with people. The bad thing about being a manager is that you are the employee of all your subordinates. The boss gets to solve all the problems that the subordinates bring, such as when they say, "I need an office."

Currently as a senior project scientist for the James Webb Space Telescope, I do some science but mostly communications and outreach. I give public lectures and talk about the progress of the project. I work with a team of 12 other scientists at Goddard to make sure that we work properly with the team of engineers that makes the project function. We need to appreciate what they do, and they need to appreciate why they are doing it.

The Space Telescope is my main job, but I do try to give some attention to trying to start new areas. I am working with one colleague at Goddard to see how we are going to follow up gravitational-wave bursts. With another, I am asking how we will make an atomic-wave interferometer to detect

John Mather had a model of the Cosmic Background Explorer (COBE) satellite in his NASA office in 2006. The satellite studied the pattern of radiation from the first few instants after the universe was formed. In 1992, the COBE team announced they had mapped the primordial hot and cold spots that were the seeds for the galaxy clusters that now span millions of light-years across the universe. (NASA photo)

gravity-wave signatures. I try to pay attention to the larger picture and look for opportunities. I enjoy that.

A SCIENTIFIC FAMILY

I have been fascinated with math and science for a long time. In first grade, I was already aware that there was such a thing as infinity: I covered my brown page of paper with a one and as many zeros as would fit on the page. My parents would read aloud to me and my sister about scientists—biographies of Galileo and Darwin. I was taught that scientists are really important. I learned that when you make discoveries, the public doesn't always like what you find, which proved how important it is to make those discoveries.

Both science and education are important to my family. My father studied the genetics and feeding of dairy cows. My mother's father was a bacteriologist who helped develop penicillin. Almost all of the members of my family who weren't scientists were teachers. My grandmothers were teachers, as were my mother and my aunts. My sister is also a teacher. So both science and education run in our family. Although my parents didn't actively encourage me to go into science, they gave me many opportunities to explore science. They found me activities that they thought I would like, such as science summer camp or the 4H Club, where I learned about electronics.

Our home was at the Rutgers University Dairy Research Station in Freehold, New Jersey, about a mile from the Appalachian Trail where it cuts across the state. It is very pretty out there, but for a kid it was very isolating because you cannot walk far enough to get anywhere. We lived on a tall hill, and it was easy to go down, but biking back up was hard and slow, so I mostly stayed up on the hill. My scientific activities were pretty much confined to the house. I was interested in electronics and astronomy. I built a radio kit with a single vacuum tube, but we were too far out in the country to get a signal. Then I built a Heathkit radio with five vacuum tubes and was able to listen to shortwave radio from all over the world. I sent away to Edmund Scientific for lenses and built cardboard-tube telescopes. By high school, I had saved up enough money to buy a four-inch mirror and a secondary mirror, and I built a real reflecting telescope.

TRYING THE IMPOSSIBLE

Throughout my career, I have tried to do near-impossible experiments. One of my first such experiments was for my high school science fair. I decided to measure the orbit of an asteroid by adding a camera and photographic film to my reflecting telescope. In 1801, Carl Gauss had determined the orbit of Ceres, which is the largest dwarf planet in the solar system, using this method. I wanted to do this too, but measuring an asteroid orbit that way is hopelessly difficult. As in the later cases when I attempted a really hard experiment, I learned a lot from this experience. It also gave me a lot of respect for Gauss. I learned early on about the near-impossible experiment that you might as well try.

I did succeed at some hard things in high school. I taught myself calculus from reading a book. I found that by doing every problem in a textbook, I could understand it. Learning from a book is a self-contained project, and the trajectory is clear, but when you are building something that you have never built before, there is no guide how to do it. All you can do is get started and empirically try to make it work; you "poke" at it. That's how I define real science—you get started and "poke" at it.

Students in school are taught that science is done by starting with a hypothesis and then designing an apparatus to test your hypothesis. I think that concept is ridiculous—that is not what real scientists do. Real scientists work directly with experiments, hands-on, and play with them. As a young person, I had gears and pulleys and motors to play with and learned about hands-on science. It bothers me that many applied scientists never had such an opportunity.

At the end of high school, I applied to six schools and was accepted into all of them. I visited Harvard and MIT, Swarthmore and Rutgers. I liked the people I met at Swarthmore. As a country kid, I was a bit intimidated by the big city of Cambridge and drawn to Swarthmore's country setting in Pennsylvania. But after a few weeks at Swarthmore, I started to wonder if I was the only physics student at the entire school. I was thinking of transferring but instead got special help from the physics faculty and jumped ahead.

As an undergraduate, I was more interested in fundamental particles and quantum mechanics than in astronomy. The Swarthmore Astronomy Department was doing conventional telescope observation, looking at the position of stars. That seemed boring. The field of cosmology was just getting started—cosmic microwave background radiation (CMBR) was discovered when I was a freshman. A young visiting professor from Princeton, Dave Wilkinson, graded my senior honors exam. He was important later to our understanding of cosmology, contributing both theoretical and experimental results. After his death in 2003, the satellite probe that was then measuring the anisotropy of the microwave background radiation and verifying our current cosmological model was renamed in his honor.

When I finished my undergraduate degree, I visited some friends who had gone to Princeton for graduate school. I was admitted and planned to go there as well. But the school seemed isolated, and there were few women enrolled there. I decided it wasn't where I wanted to be. I wrote to the Physics Department and told them, "I'm not coming to your graduate school because you don't have any women." An old friend from high school was at Berkeley as a graduate student and sent a photo of himself in a short-sleeve shirt, sitting on a fountain in January. He suggested that I come to Berkeley and get a summer job. I got a job at the Lawrence Berkeley National Laboratory working on a spark chamber. I liked Berkeley and asked the Physics Department if I could stay, and they said I could. So my choice of graduate school was not carefully planned—in fact, just the opposite: it was happenstance. Some folks might say about me that I am really smart, so I must have known exactly what I was doing.

Such was not the case.

When I first started graduate school at Berkeley, I thought I wanted to be an elementary particle physicist like my hero Richard Feynman. I took courses for two years, during which my faith in my future was being challenged by the Vietnam War and by demonstrations that rocked the city. Because I was very nearsighted, I was not drafted to be a soldier, so I didn't have to directly deal with the great issues of principle that involved so many of my classmates. For a while I considered studying the law in order to defend the country from the government of the day, but when I read the course catalog, I couldn't imagine studying those subjects. Now, because I have been a longtime government employee, my perspective has changed a bit. I also talked with my plasma physics professor about developing nuclear-fusion power for the good of humanity, but he seemed to think this would be an extremely long and difficult project, as it has turned out to be.

In 1970, I was looking for a thesis project. Professor Paul Richards was working with Charles Townes and a young postdoc named Michael Werner to study the newly discovered CMBR. I liked all three of them as well as the proposed experiment. The first project was to build an infrared spectrometer to take to the Barcroft station on White Mountain in eastern California. This project worked but was limited in accuracy by the interference of Earth's atmosphere. We set limits on the CMBR intensity and wrote a few

John Mather shared the Nobel Prize in Physics in 2006 for his work confirming the Big Bang theory of the early universe. Mather (*far left*) and his NASA colleagues (*counterclockwise starting next to Maher*) Al Kogut, Gary Hinshaw, and Chuck Bennet displayed a satellite map of the universe's microwave background radiation in October 2006, when the Nobel Prize was announced. (NASA photo)

publications. Our next experiment was a balloon-borne far infrared inter-ferometer to measure the CMBR spectrum. This was the beginning of a bap-tism by fire in the art of building instruments that would work in remote and hostile locations. I had to learn something of almost every area of engi-neering, from mechanical to optical to cryogenics to electronics. My skill was stronger in understanding than in implementation, and the antenna on the balloon payload fell off while it was on the launch pad when my sol-der joint broke. Fortunately for us, this fault was noticed, and the payload was launched successfully.

LEARNING MURPHY'S LAW

However, it also was true that we got tired of testing, and our instrument did not work, for three different reasons. The awful feeling these failures produced stayed with me for the rest of my life, and this experience was one of those ways of learning what one does not want to learn. Murphy's law had been proven one more time. Also, I wanted to finish my thesis and had already lined up a job in New York as a postdoc with Pat Thaddeus. My adviser agreed, and my thesis was limited to the ground-based work and the design for the balloon instrument. After I left for New York, my graduate school colleague David Woody figured out why the instrument had failed, fixed it, and made it work on the second flight. He analyzed and published the data and got his thesis out of the project, too.

With Pat Thaddeus at the Goddard Institute for Space Studies, a part of NASA housed in a building adjacent to Columbia University, I was hoping to go into a new field of study. I thought that my work on the background radiation was very difficult, and it was going to be hard to do much better with balloons. I arrived in New York at the end of January 1974 and started theoretical and observational work on naturally occurring silicon-monoxide masers. I learned how to build a microwave receiver with brilliant machinists and technicians, and I took it to the McDonald Observatory in Texas and to the US Navy's Maryland Point observatory on the Potomac. We did observe the silicon-monoxide emission at 43 gigahertz, which had never been seen before in space, and I made some progress writing a giant Fortran program on the IBM 360 computer, but it never came to anything, and years later I threw many boxes of IBM cards into the trash, finally admitting defeat.

In the summer after I arrived at Goddard, my trajectory took another abrupt turn when NASA issued Announcements of Opportunity for Scout- and Delta-launched satellite missions. My optimism was returning, and when Pat asked for ideas, I cheerfully asserted that my thesis experiment

John Mather is a senior astrophysicist at NASA's Goddard Space Flight Center in Greenbelt, Maryland, and a senior project scientist on the James Webb Space Telescope, which is scheduled to launch in 2020. (NASA photo)

would have worked much better in space. He suggested that I call up Rainer Weiss, David Wilkinson, and Michael Hauser, and with their colleagues Dirk Muehlner and Bob Silverberg we conceived of the new mission. It would have four instruments, a far infrared interferometer to measure the CMBR spectrum, two instruments to measure its anisotropy (difference in brightness in different directions), and an instrument to hunt for the diffuse infrared background from the first galaxies. The balloon payload flew successfully after David Woody fixed it, and things were looking up. We sent in our proposal, typed by hand on a real typewriter. I thought that the odds of our mission being selected were very low. None of us had any prior experience with space missions, and none of us knew that there would be about 150 other proposals or that two of those (from the Jet Propulsion Laboratory and Berkeley) would be direct competitors to ours.

A NASA MISSION

However, NASA was interested after all. There was already a negotiation with the Netherlands and the United Kingdom to build the Infrared Astronomical Satellite (IRAS). The first expression of interest in our idea was to see whether the spectrometer could be miniaturized and given a ride to space as some part of the IRAS. I presented my concept to the IRAS science team at a meeting near Amsterdam. It went over with a resounding thud, for good reasons. I am very glad I never had to build the first version of the instrument I had conceived, but I did learn a lot about what could be done, and I learned about the IRAS mission.

We submitted our report, and the results were favorable, so NASA assigned us a larger team of seasoned engineers. This was a time when the Space Shuttle was being considered by Congress, and NASA made a deal that would set its future for a long time: all new launches would be made with the shuttle, and all the expendable rockets such as the Deltas would be canceled. We argued but without success, and we had to redesign the COBE to go on the Space Shuttle. This wasn't so easy because the COBE needed a polar orbit, achieved by a launch from California, at around 900 kilometers altitude. Most of the shuttles would be launched from Cape Canaveral (then called Cape Kennedy), so our requirement was a challenge in any case. By around 1979, NASA decided to build the COBE satellite in-house at Goddard, meaning that engineers and scientists at Goddard would work together very closely. This is an exception to the usual way that NASA obtains satellites, which is by writing contracts to major aerospace organizations and university laboratories. In our case, two of the three principal investigators were already at Goddard, and the third (George Smoot at Berkeley) was willing to have Goddard build that instrument, so this new arrangement was very good for us. We had daily interactions with our engineering colleagues; we could walk into any laboratory to talk about any problem, and we made significant forward progress. I really enjoyed that part. But part of the deal was that our project would also be a training project for new engineers and a reservoir for talent. When other projects got into trouble, our team was raided for top talent to go solve emergencies elsewhere, and of course there were many such emergencies. I was very frustrated about this state of affairs, but I had to admit: the Hubble Space Telescope really did have priority.

In 1980, I made a major decision to marry Jane Hauser (no relation to Mike). I had met Jane in New York in 1974 while I was taking a workshop in reevaluation counseling, one of many personal-growth experiences that I sought as a part of my emotional education. (My sister Janet became a teacher of this subject for many years, as did one of my many bosses at Goddard.) Jane is a ballet teacher, but she was taking computer programming and math courses as she completed her undergraduate education, and I was very impressed. So on November 22, 1980, a hundred scientists, engineers, and dancers threw us a potluck banquet after our wedding, and I have never seen so much good feeling and good food at one time and in one place, at least until I arrived in Stockholm. Jane has been my life partner, my best friend, my best editor, and my best adviser ever since. With her, I have traveled to many amazing places and become quite fascinated with understanding how ancient civilizations managed to accomplish their engineering feats. We have seen Tycho Brahe's observatory in Denmark; we have seen Ulugh Beg's observatory in Samarkand; and, I think most amazing of all, we have seen Pompeii, with plumbing, faucets, running water, and so many signs of modern life that one can hardly imagine how that knowledge was lost. Sometimes I think it would be fun to write books about how great cities were built, but I seem to have something else to do right now.

ABOUT THE SUBJECT

John C. Mather is a senior astrophysicist at NASA's Goddard Space Flight Center in Maryland and adjunct professor of physics at the University of Maryland–College Park. He received his BS from Swarthmore College and his PhD from the University of California–Berkeley, where he was a Hertz Fellow. Mather won the Nobel Prize in Physics in 2006 for his work on the Cosmic Background Explorer satellite, which measured the black-body spectrum and anisotropy of the cosmic microwave background radiation.

36 QUIET SCIENCE AND NUCLEAR THREATS

Jay Davis

The Iraqi security forces had guns; my United Nations (UN) inspection team wasn't allowed cameras or secure radios, let alone weapons. But in June 1991, when we stumbled upon 100 trucks' worth of contraband nuclear weapons equipment, the only weapon we needed were radios broadcasting in the clear. While the Iraqi guards shot at us, we called for help over wavelengths we knew would be picked up by local media and then spread internationally.

The resulting media firestorm and the United States making clear that it would be happy to restart a shooting war were all we needed to get Iraq to declare its nuclear program. In fact, the regime revealed not only that it was using the electromagnetic separation methods we already knew about but also that it had been experimenting with centrifuges with advanced, carbon-fiber rotors.

Nothing can really prepare you for getting shot at, and it certainly wasn't what I expected when I entered the field of physics, but I was well positioned for a job on the UN team looking for evidence of a clandestine nuclear program. In my day job at Lawrence Livermore National Laboratory, I had built electromagnetic separators (albeit for analytical purposes rather than military ones), so I was well acquainted with the equipment, shielding, and human infrastructure needed for a nuclear research facility.

As proud as I am of my work in Iraq, I am perhaps prouder of my quieter work building up the research facilities at Livermore. The ultrasensitive mass spectrometers and the electrostatic accelerators I helped design opened new fields in materials science and biology, provided the training ground for a new generation of scientific leaders, and inspired similar designs in more than 100 facilities worldwide.

I have always thought of my leadership style in my work in Iraq, at Livermore, and later with the Hertz Foundation and as the first director of the Defense Threat Reduction Agency as "environmentalist": not in a "green" sense, but in the sense that my role was to create the teams and environments where people could succeed on their own merits, working together for the success of both individuals and the team. That approach has served me well in jobs around the world, whether I was managing particle physicists or soldiers.

SCIENCE AND DRAMA CLUB

I was born in Texas in and moved with my family to Austin shortly thereafter. I enjoyed math and reading. Although I never had a good English teacher, science was omnipresent in Texas in the 1950s. I was a high school sophomore when Sputnik was launched, and I spent every summer at science camps. It was never a question of whether I would have a career in science. The question was which scientific field to enter: astrophysics or engineering—space or the rockets to get us there.

Despite my scientific abilities and interest, I frustrated my science teachers by refusing to participate in after-school science activities. Instead, I spent my time in drama club. I wasn't particularly talented, but some of the acting pointers would come in handy years later in Iraq. Between that and my childhood in the Methodist Church, where I was fortunate enough to hear preachers who really knew how to use words, my public-speaking skills grew enough to serve me well my whole life.

I received an excellent high school education, mostly from female teachers with master's degrees who would have been practicing researchers in a later era. The University of Texas–Austin, a superb university that happened to be the local school, was the obvious choice for college. Despite my desire to be a scientist, I found myself coasting through my course work, depending mainly on German classes to keep my grades up. This lasted until my junior year, when I realized if I didn't improve my physics grades, I wouldn't be able to pursue what I cared about. Once I buckled down, I was able to graduate in three years, with a BA in math and a BS in physics. While in college, I married my high school sweetheart, Mary, and we have been together since.

Jay Davis married his high school sweetheart, Mary, while in college. He started graduate school in physics at the University of Wisconsin—Madison, shortly thereafter, in 1964.

LEARNING NUCLEAR PHYSICS

In 1964, at age 20, I was done with college, married, and on my way to the University of Wisconsin–Madison to get a PhD in physics. I chose Wisconsin after becoming friends with Harold Hansen, the head of the Physics Department at the University of Texas, who had gotten his own PhD at Wisconsin and recommended it. Many of the professors at Wisconsin were alumni of the Manhattan Project. They had returned from Los Alamos with all manner of accelerators and other equipment and had set to work trying to understand the parts of nuclear physics that weren't needed to build the first atomic bomb.

My adviser was Heinz Barschall, who had a fascinating life. Raised Lutheran in Germany, he was forced to leave the country in 1937, after attempting to enroll at the University of Berlin and finding out he was disqualified by his Jewish ancestry. He fled to the United States, where he aced the entrance exam for Princeton's physics program despite needing an

English–German dictionary to translate the questions. After completing his PhD at Princeton, where he was among the first to fission uranium with fast neutrons, he taught physics at Kansas State. There, as an "enemy alien" during World War II, he was under so much suspicion that he wasn't allowed to continue nuclear research, even while being carefully watched. His status changed when the Manhattan Project needed his expertise with fast neutrons; he was quickly granted citizenship—one of only a few Germans to receive citizenship during the war—and put on a train to New Mexico.

When I met him, years after the Manhattan Project was completed, Heinz was a dour personality but a wonderful teacher, and we bonded quickly. He taught me to control my occasionally overwhelming extraversion, and the writing I did under him vastly improved my command of the written language.

It helped that I was enthusiastic about the work I was doing with him— measuring the cross-sections of fast neutrons at energies between 0.5 and 30 megaelectron volts. For me, the physics of the project were secondary to the engineering problems—I was realizing that I was a gifted technologist. I'm most in my element when I'm conceiving of and building complicated instruments and when I'm building teams to do the same.

Unlike the military-run Manhattan Project, Heinz's work at Wisconsin was pure science, measuring the fundamental properties of fast neutrons. But with the intensifying Vietnam War, those were tense times at universities across the United States, and Wisconsin was no exception.

In August 1970, a group of antiwar partisans calling themselves the New Year's Gang car-bombed the building where Heinz's lab was located. Their target was the Math Research Center, a US Army–funded think tank on the second to fourth floors of the building. The Physics Department, however, bore the brunt of the attack. A friend of mine, Rob Fassnacht, was killed. Heinz's 24-year-old lab was completely destroyed. My last nine months in Wisconsin were spent rebuilding the lab and helping Heinz recover. He eventually swore off nuclear physics and took a leave of absence at the Lawrence Livermore National Laboratory to begin a program in neutron acceleration for biomedical purposes. I soon followed.

MANAGING PROJECTS

At Livermore, I managed a few smaller projects and developed a reputation as a good manager. That turned out not necessarily to be a good thing because I was then assigned to clean up troubled projects. I usually found problems on these projects to be due to human issues, not to technical ones. For example, on one project the head of the program was a theoretical physicist who was micromanaging engineering decisions he wasn't qualified to make. As a result, large orders to vendors were cancelled at the last minute, losing the lab the opportunity even to get refunds. I tried to point out the failures, but I wasn't high enough in the program to fix the human problems.

I was ready to leave Livermore when the director gave me a chance to help take the lab in a new direction. Until the 1980s, Livermore had been supported primarily through government-funded weapons research. I thought about ways that Livermore could support other research goals that were consistent with our expertise. I had experience in accelerators and so looked for an accelerator research tool that would let scientists using our facilities succeed in a new area of science.

The tool I settled upon was accelerator mass spectroscopy, in which an electrostatic particle accelerator—a giant van de Graaff generator—would accelerate ions through a magnetic field to identify which atomic isotopes they contain. I knew this tool could accelerate research that required isotopic analysis, such as radiocarbon dating of archaeological samples or isotopic analysis of ecosystem samples, but I didn't realize how dramatically the tool would change the biomedical sciences.

Building this new tool was a big investment for the lab, and I had to cajole my managers into taking the risk. Fortunately, I could take some shortcuts to reduce costs. The electrostatic accelerator didn't require the bulky and expensive concrete shielding of many Livermore projects. The most expensive part of the instrument should have been the accelerator itself, but by collaborating with the University of Washington and purchasing their used accelerator, we were able to halve the price of the overall project.

Because the accelerator could take liquid samples, for the first time extremely sensitive mass spectroscopy could be used on biological systems. A biopsy—a tissue sample from a lab animal—could be loaded into the instrument, and its isotopes analyzed. This allowed biologists to trace the path of miniscule quantities of molecules through organisms' bodies, merely by tagging them with heavy isotopes such as carbon-14. After being given a microdose of a drug, a mouse could be sacrificed, and its DNA could be extracted to find where and how much the carbon-14-tagged drug was binding to its DNA.

More than 200 graduate students have done research for their dissertations using the accelerator facility at Livermore. Now that the value of accelerator mass spectroscopy for biological investigation has been proven, hospitals and pharmaceutical firms are also building their own accelerators.

Working on isotopic separation also introduced me to the world of nuclear enrichment, the separation of radioactive isotopes required for producing

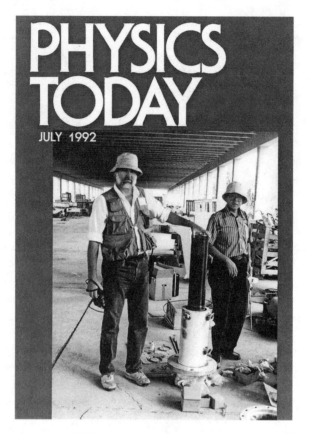

Jay Davis managed the UN inspection team searching for contraband Iraqi nuclear weapons equipment. His picture was on the cover of *Physics Today* after reporting team findings to the UN Security Council in 1991.

enriched uranium for nuclear fuel and for nuclear weapons. That introduction led to my volunteering with the Nuclear Emergency Search Team (NEST) and becoming trained to search for terrorist nuclear devices. That training put me on the shortlist for US nonproliferation efforts.

INSPECTIONS IN IRAQ

It was on a Friday that I got the call from the lab head for UN inspections in Iraq.

"We need a mass-spec guy for the next inspection in Iraq," he said. "Can you be in Washington Tuesday and then on to Bahrain by Thursday to join the team forming there? Just go out, buy what you need, and throw the receipts back through the door on your way out of town."

The sort of improvisational skills you pick up from drama club and engineering management came in handy when I started stocking up for my trip. A quick shopping trip gave me most of the tools I would need, but missing were the swipes, labels, and baggies for a serious sampling campaign of all those labs in Iraq. Fortunately, I remembered our daughter Kathy's archaeology firm in Santa Cruz. After pillaging the firm's storeroom, I promised the firm two days of free access to Livermore's accelerator for radiocarbon dating.

The improvisation continued when I landed in Baghdad. Our Egyptian interpreter, assigned to handle conversations between my team lead and his Iraqi counterpart, was far more accustomed to political discussions than scientific ones. He was concerned that his English vocabulary didn't include the words for the technical physics, chemistry, and engineering discussions we would have.

In a moment of inspiration, I taught him a cramming trick I wish I had known in university. I handed him my textbook on electromagnetic isotope separation and told him to just read the figure captions. If it's important, it'll be in there. The next morning he came down to breakfast with a huge smile on his face and threw his arms around me. "I've got it!" he said, and he did. He may not have had the skills to build an electromagnetic separator, but, then, I don't either—that's what your team is for.

My physics and counterterrorism training may have been what brought me to Iraq, but I ended up using every tool in my kit to do my job there. It was a thrilling hunt looking for evidence of the nuclear program, made even more so because we were unarmed and constantly face-to-face with armed guards. My public-speaking training—and height—gave me the confidence to keep control over some very tough situations, perhaps more than I realized at the time—. The chairman of the Iraqi Atomic Energy Commission told me afterward that they all thought I was some sort of CIA thug. I even found a few moments to throw around the German I had learned years ago, enough to keep my counterparts on their toes as they wondered what other languages I knew.

Managing my own team could be as hard as managing the Iraqis. When we stumbled upon that convoy of nuclear equipment out of Fallujah, some of the team performed beautifully, calling for help on the radio to create an international incident and stuffing camera film in their underwear to keep our documentation of the event safe. Others began to scream that David Kay, our team lead, had created an international incident and that we must

Jay Davis helped found the Center for Accelerator Mass Spectroscopy (CAMS) at the Lawrence Livermore National Laboratory in 1988. CAMS accelerators have had broad impact on a range of chemical, biological, geological, and materials problems. Jay stands in front of the largest CAMS accelerator, circa 1993.

disavow him and leave immediately. David ordered me to put the troublesome team members forcibly back on our bus to keep them out of his hair while he managed the crisis. I enjoyed the task.

My proudest bit of information gathering wasn't subterfuge or coolness under fire but simply talking to the Iraqi scientists as scientists. One day, after weeks in which we tried to talk to worried scientists who clammed up whenever their security minders were around, the minders finally wandered off to follow the chemistry team. I told the Iraqi scientists that I would just draw on the chalkboard what I thought they had done, and they could correct me when I was wrong.

For the next hour, the Iraqis cooperated completely. Every time I got a small detail wrong (occasionally by intention), they would rush to correct me. They were very good scientists, and as time passed, the question-and-answer session became instead a discussion among peers. The Iraqi scientists and engineers were overjoyed to describe their clever designs.

RETURN TO THE WEST

After we helped uncover the Iraqi nuclear program and then briefed the UN Security Council and the International Atomic Energy Association on it, it took a while for the media attention to die down. Then in 1997 the US Department of Defense created the Defense Threat Reduction Agency (DTRA) to consolidate several programs and centralize nonproliferation activities. I was asked to serve as the first director and jumped at the chance. It wasn't easy to merge previously independent departments into a new government organization because each group in such a merger believes it is getting the worst of the deal. But I am pleased that at DTRA I was able to create an environment where the civilian and military personnel could work well together and be effective.

After I retired from DTRA, I managed another organizational shift as president of the Hertz Foundation, overseeing its transition from a private charity to a public foundation. At both DTRA and Hertz, I was happy to be given the autonomy needed to create an environment for success.

ABOUT THE SUBJECT

Jay Davis is President and Director Emeritus of the Fannie & John Hertz Foundation. He is a nuclear physicist with an undergraduate degree from the University of Texas–Austin and a PhD from the University of Wisconsin–Madison.

IV SUMMARY

LIFE AND CAREER LESSONS: LEARNING FROM LIVES, VIEWS, AND STORIES

"Irresistibly compelling, and I love it." That is how *Sandra Glucksmann* views start-up companies. After she starts one and helps it grow, it's time to do it again.

"When I got to the lab, it was fun—every minute of it!" That is *Rai Weiss's* description of a career of experimental work in physics and cosmology. Rai loves to build equipment, make it work, and spend time with students.

"This is high-tech stuff that looks like something out of a science-fiction movie. . . . I feel as if I am making history." That is how *Renee Horton* views her work at the NASA Michaud facility in New Orleans, where she is in charge of welding huge rocket motors. Renee always wanted to be a scientist. It required determination and a sense of purpose to overcome obstacles and claim her dream.

Entrepreneur, academic, and public servant. Three role-model scientists with different backgrounds, career paths, and stories but a shared passion for their life's work.

We can learn many lessons from these role models' lives, views, and stories. A place to start is their excitement. Role-model scientists approach the world as a fascinating place, one full of potential. With their enthusiasm, they create companies, make discoveries, and contribute to US security and competitiveness.

But the road they choose is not always straight or the direction obvious. In his essay "A Change of Direction," *Michael O'Hanlon* speaks for many of his colleagues when he says, "We write our résumés listing our careers as if they were preordained, with excellence and clarity of purpose. Sometimes I feel like writing an alternate résumé that would list all the ways I struggled and still do."

One fruitful area for learning is to notice how role models deal with career decisions during and shortly after their training. In general, they are open to making a significant change of direction. As students, they may need to change projects, choose a different academic adviser, or move into a different specialty. As professionals, many have to choose between the satisfaction of solving technical problems on their own and the increased impact of managing others to do so. Some decide to leave a secure industrial job to start a new company; others realize their work would be more meaningful to them if they became a government employee.

We can also learn by examining how the subjects of the chapters in this book balanced professional interests with family needs or other pursuits. Some of the scientists decided to work part-time to care for their families or to find jobs that allowed both themselves and their spouses to be able to pursue their careers. Some found jobs that provided time for them to actively pursue interests outside of science and technology.

Some of the women scientists describe experiences of gender bias. Similarly, scientists who are members of minority communities describe the experience of being unique—the only person of color in the room. Understanding their experiences helps us recognize bias and discrimination and suggests a path toward providing a welcoming environment for women and minority colleagues. Some subjects have worked to challenge discrimination or institutional barriers, and their actions provide a template to follow.

THE ENTREPRENEURS

The entrepreneurs made their decisions to work in a commercial setting for a variety of reasons. For some, their entrepreneurial direction began in childhood. *Richard Post* grew up in Silicon Valley watching role models such as William Hewlett and David Packard create electronics companies. Dick came to believe that what one did in life was to get a PhD, teach at a university, learn some great technology, and start a business. *Daniel Theobald* also grew up in Silicon Valley and absorbed the entrepreneurial culture. For *Kathy Gisser,* watching her parents run their own business taught her to value an entrepreneurial spirit before she was old enough to know what that meant.

Some of the entrepreneurs had experiences in high school that indicated an interest in commerce. *Stephen Fantone* built a telescope using money earned cutting grass and shoveling snow. His interest in business continued as an undergraduate at MIT, where he sold optical equipment. *Christopher Loose* had a high school internship at Merck & Co. that helped him choose a career in chemical engineering, biology, and business.

Some of the entrepreneurs were planning to go into academia or to work for the government, but events or experiences sent them in an unexpected direction. *Ethan Pearlstein* had always wanted to be a professor and do research in an academic environment. But after multiple rejections, he took an unusual approach. By founding a company to study rare diseases, Ethan created an academic research environment outside of academia. In my essay, I describe how cuts in government funding caused me to rethink my plan to pursue fusion-energy development at a National Laboratory and instead decide to go into technology development in a commercial setting.

The desire to seek a wider range of experience and to apply their expertise to practical problems led several of the scientists to work in the commercial sector. *Beth Reid* was considering an academic career but grew tired of modeling galaxy clustering. Her wide range of interests was already clear in graduate school, where she considered a thesis in ecology or biophysics before settling on cosmology. As a mathematical modeler, she developed computation and analytical expertise that could be applied in many settings. Now working at the *Climate Corporation*, she uses her modeling tools to analyze satellite data and help farmers increase crop yield. *Jennifer Park* worked as an individual contributor in a biology laboratory during graduate school and at her first job. She was interested in working on a wider variety of projects and decided that she could be more effective as the leader of a commercial team than as a lab scientist, so she moved into business development as a customer product manager.

Working on a team has also been important to *Sandra Glucksmann* and *Z. Jane Wang*. Sandra really appreciated her first team at a start-up biotech firm, where her colleagues complemented and supported each other while working toward an ambitious goal. After her postdoc, Jane had the choice of continuing at a start-up that she and coworkers had founded or move to Verily, Google's life sciences unit. Jane chose to become a technical leader on a large team, where specialists combine their expertise to tackle multidimensional biological and computational problems.

With an interest in system dynamics and a cross-disciplinary ability to understand how large systems function, *Wanda Austin* and *Norman Augustine* were highly valued technical contributors at their aerospace companies. Both also showed an interest in the commercial side of the defense business: Wanda studied US Air Force contracts, and Norman joined the US Department of Defense as a program manager. With their technical and business understanding, they moved up through management to become CEOs. And each learned early on the value of teamwork and developed the managerial skills needed to lead teams.

Collaboration and teams became more important to *David Galas* as he transitioned from theoretical physicist to experimental biologist. David had always enjoyed solving mathematical puzzles on his own. In graduate school, he was beginning to see the advantage of collaboration for analyzing complex physics problems. Later as the manager of the US Department of Defense Genome Project and the founder of biotech companies, he considered teamwork crucial to making progress. He also acquired the managerial skills needed to lead large teams.

David is an entrepreneur who has spent time in all three sectors: academic, commercial, and government service. He comments on the advantages each setting has to pursue important science: there are passion-driven, peculiar, creative projects you can do only in academia; massive endeavors that require government backing; and rapid, focused, and applied team efforts best pursued at a start-up company. He says that just as there is no single pathway from graduate student to successful scientist, there is no single pathway from basic research to technological innovation—in each sphere, different stages and methodologies inform each other in surprising and productive ways.

THE ACADEMICS

The academic scientists took various routes on the way to a career as researcher and teacher. Some were influenced by having parents who were college professors. *David Spergel's* father had a satisfying career as a physics professor, mostly teaching students who were the first in their family to go to college. *Tamara Doering* had two academics as parents. She says that she grew up believing that getting a PhD and teaching at a university were the normal progression for every student, the job that everyone has after finishing school. *Richard Miles* had a parent who was a research engineer and a grandfather who was a professor. For Richard, becoming a research professor in mechanical engineering was a natural path.

Some of the academic scientists learned during college that they enjoy teaching. *Jessica Seeliger* was a tutor in college, and *Kathleen Fisher* was a section leader who taught computer science to other undergraduates. Some academics learned about the teaching profession after college. *Shirley Tilghman* and *Michael O'Hanlon* joined service organizations to teach science in foreign countries. Shirley went to Sierra Leone with the Canadian University Services Overseas to teach high school chemistry. Michael joined the Peace Corps and taught physics in Zaire. *Stephon Alexander* spent part of his postdoc years as a volunteer teaching at a theoretical physics summer program. In addition to enjoying his teaching, he also found it helped him to review technical areas that he wanted to understand better.

Most of the academics chose their scientific field while in college. *Stephon Alexander* and *William Press* studied physics in college and ended up becoming physics professors. *David Spergel* studied astrophysics and learned as an undergraduate that he was better suited to be an astrophysical theorist than either a mathematician or experimentalist. *Kathleen Fisher* fell in love with computer science during her sophomore year of college and chose that as her career field.

Some academics changed scientific field after college. Both *Jessica Seeliger* and *Shirley Tilghman* majored in chemistry in college and later became biologists. In both cases, they sought out mentors and environments that would allow them to acquire the additional biology training they did not receive as undergraduates. For Jessica, that meant going to England for a master's degree in protein folding under Alan Fersht before starting PhD work in biology. For Shirley, that meant postdoc training at the National Institutes of Health (NIH) in the lab of molecular biologist Phil Leder.

The academics found during graduate school and postdoc work that they enjoyed the process of becoming experts in specialized areas. *Tamara Doering* fondly remembers the day when in a flash of inspiration she decided to work on the pathogenic fungus that she has now been studying for more than 20 years. *Jessica Seeliger* continues to study fundamental aspects of the tuberculosis bacteria, most recently examining the microbe's lipid transport, cell walls, and membranes. *Richard Miles* became an expert at using lasers to study aerodynamics and combustion. *Shirley Tilghman* ignored advice from colleagues and spent years studying the H19 gene. It was a puzzle that she wanted to solve. She did solve it, and the H19 gene turned out to be the first long noncoding RNA to be described in biology. *Rai Weiss* spent 40 years improving apparatus designed to detect gravitational waves, with no guarantee such waves would ever be seen.

The academic scientists have found collaboration with colleagues, both inside and outside their research groups, to be important for productivity and creativity. As a young professor at Penn State, *Stephon Alexander* learned the advantage of talking not just to cosmologists but also to condensed-matter theorists. Stephon was able to use some condensed-matter concepts to model the early universe. Astrophysical theorist *David Spergel* has made headway in instrument design by collaborating with experimentalists and engineers. Their joint efforts have recently resulted in a number of papers on better ways to detect exoplanets. *Shirley Tilghman* recounts how as a postdoc at the NIH she and another researcher were given the same assignment, which could have been a recipe for disaster. But they instead divided up the jobs, worked well together, and made an important discovery about how mammalian genes are organized. Their joint discovery propelled both researchers into their independent scientific careers. *William Press* collaborated with three of his colleagues to write the widely used scientific computing textbook *Numerical Recipes*. And as one of the founders of the Laser Interferometer Gravitational-Wave Observatory scientific collaboration, *Rai Weiss* has more than 1,000 colleagues who contribute to the observatory's work.

Most of the academics became professors after graduate school or after completing postdocs. *Kathleen Fisher* and *Michael O'Hanlon* are exceptions. Kathleen spent a full career in industry before returning to academia to teach. Michael O'Hanlon worked first in government, then joined the Brookings Institute to do policy research, and is currently an adjunct professor at four universities.

Most of the academic scientists have stayed at the same institution and worked in the same general scientific field for much of their careers. *William Press* is an exception and says that he has changed fields approximately every five years, even when staying at the same institution. After writing a thesis in relativity, he moved into cosmology, then into fluid dynamics, and most recently into molecular biology. At one point, he also spent time directing weapons research at Los Alamos National Laboratory. A common thread across these fields was his expertise in scientific computing.

Like *Bill Press*, *Shirley Tilghman* also changed scientific fields to become a biologist, but she did so much earlier in her career. Shirley changed fields in response to a comment by a chemistry professor who told her that she was a good student but wasn't going to make a good chemist. She knew the professor was right and considered this advice to be kind and honest. She has made this same suggestion to students a couple of times, but she says she has to be 100 percent certain to offer such advice because it can have a serious impact on a student. She also says that she is very careful about how to deliver the message.

According to Shirley, knowing a change is necessary and admitting that you are going down the wrong path take self-awareness as well as self-confidence. Some of the symptoms can be unhappiness or boredom—in Shirley's case, it was the latter. She just wasn't that excited about becoming a chemist. She missed the "thrill of the chase," which she remembered from science studies in high school, when she considered chemistry an amazing way to solve puzzles.

Bill Press also has advice for students and scientists who are interested in a change of career field. He says that a person who is broadly interested in new things will be fighting a battle to move into new areas and should not do so superficially. He suggests collaborating with people who are experts in the new field and pursuing problems they find interesting. As a counter-example, he cites physicists who have moved into biology but who work on problems that the biologists think are physics. He finds nothing wrong with that but doesn't consider it a true change of career.

Bill is now a computational biologist at the University of Texas. His computational graduate students spend half their time in the lab with biology graduate students doing experiments. He encourages computational students to be intimately involved in the original gathering of data to understand the limitations of the data and how they can be made better for sophisticated analyses. Bill's comments mirror those of *David Galas*, who also made a transition from physicist to biologist. Like Bill, David also found that hands-on laboratory experience was necessary for a theoretician to understand the minute details of data generating in the lab.

THE PUBLIC SERVANTS

These scientists took various paths on the way to a career with the military or at one of the government agencies. For some, the path began in childhood. *Renee Horton* had wanted to work for NASA since middle school, when she decided she wanted to be an astronaut. *Ellen Stofan's* father was a rocket scientist for NASA, and her mother was a science teacher. Ellen was only four when her family traveled to Kennedy Space Center to see a rocket launch.

Some in this group had internships at government laboratories during college. *Jennifer Roberts* spent one summer doing research at Johns Hopkins Applied Research Laboratory and another summer at the Naval Research Laboratory. *Jami Valentine* spent summers at the Lawrence Livermore National Laboratory (LLNL).

Two in this group ended up joining the military at the beginning of college. *Paul Nielsen* chose to attend the US Air Force Academy rather than MIT because finances mattered to his family. *Ellen Pawlikowski* decided to join the Reserve Officers' Training Corps (ROTC) during her freshman year of college mainly because she was curious about the military at the end of the Vietnam War.

One of the scientists pursued her PhD at a National Laboratory. *Kimberly Budil* worked on lasers and detectors at LLNL as a graduate student in the Department of Applied Science at the University of California–Davis.

Several found jobs at National Laboratories and government agencies immediately after finishing their PhD theses elsewhere. *Jami Valentine* joined the US Patent Office after completing a solid-state physics and materials science thesis at Johns Hopkins. *Wendy Cieslak* and her husband, Mike, graduated as metallurgists from Rensselaer Polytechnic Institute and felt lucky that they both were offered jobs at Sandia National Laboratories in Albuquerque, New Mexico. *Ellen Stofan* joined the Jet Propulsion Laboratory initially as a postdoc on the *Magellan* Venus mission after finishing her PhD at Brown. *John Mather* joined NASA as a postdoc at the Goddard Institute for Space Studies after completing balloon cosmology experiments at Berkeley. *Jay Davis* followed his thesis adviser to LLNL after their laboratory at the University of Wisconsin was destroyed in a Vietnam War–era terrorist bombing.

Two of the scientists worked for a short time in industry before working for the government. *Jennifer Roberts* worked at a defense contractor on Defense Advanced Research Projects Agency programs for several years before becoming a program manager for the agency. *Deirdre Olynick* spent two years working for semiconductor equipment companies before finding a job in nanofabrication at the Lawrence Berkeley National Laboratory.

Upon joining the military or government agencies, many of the scientists started as individual contributors but quickly moved into management. After *Paul Nielsen* completed his PhD in plasma physics, he joined the US Air Force Special Projects Office to design spacecraft. He considers that job to have been the high point of his technical career. After Paul left the Special Projects Office, he was assigned managerial responsibilities. In her first air force assignment, *Ellen Pawlikowski* worked in a lab responsible for monitoring the Nuclear Test Ban Treaty. She had a team of people helping her build separation apparatuses. She found it very rewarding to empower her team to carry out the details. Both Paul and Ellen continued to be promoted, commanding ever-larger teams of air force personnel.

Both *Wendy Cieslak* and *Jay Davis* started as experimental scientists but were promoted and over time joined the management at the National Laboratory where they worked—Wendy as a director at both Sandia and Los Alamos Labs and Jay as a director at LLNL before moving to Washington to be the first director of the Defense Threat Reduction Agency.

Kim Budil worked on laser research at LLNL for several years after her PhD and before moving to Washington, DC, to help advise the National Nuclear Security Administration. Working in Washington meant stepping away from research and taking on a portfolio of projects in which she was not an expert. Kim learned that her background made her one of the most technically qualified in the Washington policy environment, and she found that she enjoyed explaining technical subjects to nontechnical people. She learned that she was interested in many different scientific topics and that she was a strategic thinker and good at motivating people. Understanding her strengths helped her take a leap and transition into management.

After Kim returned to LLNL, she rose through the ranks of management. She is currently responsible for the University of California's oversight of three National Laboratories funded by the US Department of Energy. Kim describes much of the research performed at these laboratories as medium-term projects in the areas of national security, energy, and biotechnology. The projects require several years of work by teams of qualified technologists to solve difficult technical problems. Kim says that the type of research best performed at National Laboratories is longer term than the research-and-development projects that industry is willing to fund but more applied than the basic research often pursued in academia.

Kim believes that choosing to work in industry, academia, or a government research laboratory depends on a scientist's preferences for project timescale and degree of application. Individuals who prefer to create commercial projects based on shorter-term research are more comfortable working in industrial settings, whereas individuals who devote their careers to delving deeply into basic research are more comfortable in academia.

The life stories told by the scientists and engineers in this project support Kim's view. Corporate scientists *Kathy Gisser* and *Jane Wang* use their chemistry and biology skills to make products, whereas academician *Shirley Tilghman* uses her understanding of chemistry and biology to solve puzzles of nature. Business manager *Jennifer Park* enjoys working with customers on a variety of projects, whereas for *Tamara Doering* studying one particular fungus is the focus of her scientific career. Entrepreneur *Sandra Glucksmann* grew up in an atmosphere of constant change. She embraces the uncertainty that accompanies start-up ventures and can't wait to get a new one going, whereas *Stephon Alexander* likes pursuing long-term basic research creating mathematical models of the early universe.

Kim Budil's description of scientific work at the National Laboratories is also consistent with the stories of the scientists in this project. *Wendy Cieslak's* battery-development work and *Renee Horton's* welding of large rocket motors are good examples of applied science on topics of national importance, as are the military science projects that *Paul Nielsen* and *Ellen Pawlikowski* have managed during their careers.

Questions to consider: What timescale of project am I most comfortable pursuing? Am I more interested in creating products or understanding the natural world? How do I define service to the greater good, and does that guide me toward pursuing one of these career paths?

GENDER BIAS AND DISCRIMINATION

Many of the women scientists encountered various degrees of gender bias or discrimination during their careers. *Deirdre Olynick* experienced clear and overt sexism as an undergraduate at North Carolina State. She remembers a professor looking up girls' skirts and another asking whether a prospective female applicant was pretty before deciding whether to hire her. Such attitudes made Deirdre more determined to succeed as a scientist.

Shirley Tilghman remembers hearing sexist remarks by a physics professor at Queen's University. She thought to herself that the guy was a jerk rather than thinking that maybe he was right, and girls shouldn't be physicists. She also recalls discriminatory remarks made to her near the end of her NIH postdoc, when both she and a male colleague were looking for jobs. Shirley feels lucky that such comments were not common during her training, and the two mentors whose opinions she respected were saying the opposite. Whenever appropriate, her mentors told her that she was doing well, that she should keep going and not to let anyone tell her differently. Shirley realizes many of her female colleagues faced more discrimination than she did. She thanks her parents for giving her the gift of self-confidence—the sense that no one could tell her what she can and cannot do with her life.

During graduate school, *Ellen Stofan* was conscious that there were very few women in her field, which made her feel as if she had to work twice as hard to be taken half as seriously as the men. Later, when she was working on the *Magellan* mission at the Jet Propulsion Laboratory, she sometimes felt that people seemed impressed that she could string two sentences together. Her colleagues weren't rude, but they were condescending and didn't take her seriously because she was a woman.

Ellen Stofan commented on the additional stresses that women and minorities feel when they are acting as representatives of their perspective groups. When she didn't see anyone that looked like her in the room, this realization preyed on her fears and made her ask, "Should I be here?" She said that she didn't have much self-confidence for years and had to act and pretend because there was no alternative. It took energy pretending to be someone she was not and constantly worrying if someone were going to figure out that she was not as sure of herself as she was acting. Ellen says that some people are naturally self-confident, but for a woman in a scientific workplace, it helps to act that way.

Wendy Cieslak experienced biased supervisors when she first joined Sandia National Laboratories. Even though Wendy was achieving results on a par with the men in her group, her supervisor refused to give her a good review. When asked why, he told her that she shouldn't care how much money she was earning because her husband was making plenty. For Wendy, that was the first but not the last time she transferred to get away from a biased supervisor. A later supervisor refused to nominate her for a promotion because he and his wife decided that a family with two young children couldn't handle two professional parents. Wendy filed a complaint with the Equal Employment Opportunity/Affirmative Action Office, which made one of her supervisors so angry that he warned her if she filed such a complaint again, he would end her career at the lab. She considered reporting him for this comment but decided not to risk retaliation.

Early in her career Wendy would tell other women facing discrimination to keep their heads down because in her case reporting discrimination had not been helpful. More recently, however, the lesson she takes from her experiences with gender bias is the importance of standing up for what is right. Even so, Wendy doesn't have specific advice to women scientists that would fit every possible situation involving harassment or gender bias. Her clearest advice is the importance of keeping careful records. That way, if there is a need to respond or if a pattern of bias becomes clear, it is possible to take forceful and appropriate action.

When Wendy started at Sandia, she didn't have colleagues or mentors who could vouch for her and corroborate her story, so she now tries to play that role for other women. She considers that sort of support to be very important because it not only helps women navigate the complaint process but also preempts any attempt at retaliation. (Although federal sexual harassment law deems such retaliation illegal, the threat of it continues to impede many women's progress across industries.)

Beth Reid says that compared to many women scientists she has been less affected by social pressures, gender bias, and overt discrimination and tended not to notice such situations affecting others. She says she learned only years later about discrimination occurring while she was in graduate school and during her postdoc. Had she had been more attuned to the issue, she might have been more helpful to her colleagues, she says.

CHALLENGES FOR SCIENTISTS OF COLOR

Scientists of color face unique challenges during their training and careers. *Wanda Austin, Jami Valentine, and Renee Horton* comment on the stress of being a highly visible minority, often the only person of color in the room.

Wanda Austin was one of only 20 African Americans on her campus at Franklin and Marshall College. As someone from the inner city and a geek, she says there were multiple ways that she could have felt isolated. She found a supportive faculty member and spent time with his family, which made her feel welcome and helped integrate her into the school community.

After college, in her first job in the white, male-dominated field of aerospace engineering, Wanda faced outright hostility from her coworkers. She switched companies and found her dream job at The Aerospace Corporation, an accepting place where she could prosper.

When *Jami Valentine* was deciding where to go to graduate school, her college mentors asked her the same questions that she still asks the students she advises: Do you want to be the first student of color in the department? Do you want to be the one who breaks the glass ceiling? Or would you rather go to a place that has seen black physicists before and you won't be an oddity or a unicorn?

Jami Valentine chose to attend Brown for graduate school, where she was one of only four or five African American students in the physical sciences. She found attending Brown to be a difficult experience. She finished graduate school at Johns Hopkins and was much more comfortable at an institution where there were more African American students and where 80 percent of the students were studying science and engineering.

Stephon Alexander grew up in the inner city, where most of his friends were ending up in jail or other bad situations. He had never even considered going to college until a high school mentor made the suggestion to him. Stephon has also felt the stress of being an outsider. He says the isolation was sociological, relating to his background and his place in society. Rather than being a member of the club, he has turned his outsider status into an advantage, allowing him to take risks without worrying that colleagues will laugh at him for taking an unlikely approach to solving physics problems.

Renee Horton experienced multiple types of discrimination at three universities during her graduate school career. When she was studying physics at Southern University, the faculty had no qualms about telling the few women scientists there that they belonged at home and not in the Physics Department. One foreign-born professor at Louisiana State told Renee that she was an idiot and didn't deserve to be in graduate school. Meanwhile, Renee, whose hearing disability was known to the university, was carefully recording and transcribing his lectures and using her two young boys to help her decipher what he was saying. She worked in a lab at the University of Alabama for five years before her professor sent her an email saying she wasn't making enough progress. After calling up several of her peers, she found out that she wasn't alone. Others were also having difficulties with clueless advisers and a lack of institutional guidance to help students make their way through graduate school. Renee finally found a supportive, fair adviser with whom she finished her thesis.

ADDRESSING ORGANIZATIONAL BARRIERS

Jennifer Roberts found organizational barriers, including confusing policies and a lack of support for students, when she entered graduate school in the Electric Engineering and Computer Science Department at MIT. The challenges she faced made her consider changing fields or leaving graduate school, but she instead adapted to grad school by attacking challenges in a way that was empowering. She first helped the department revamp its orientation for new grad students. She then became the president of the department's Graduate Student Association, in which one of her roles was to help the department understand why female graduate students were leaving the department at a higher rate than male students were. She found out that the women who were staying were not happier than the ones leaving: they just refused to quit when confronted with organizational barriers!

Jennifer and her peers worked to make their department a friendlier place, sponsoring programs on group dynamics, career options, and family work–life balance. At one of their workshops, women students brainstormed on ways to respond to negative comments and critiques from students and professors. They also discussed thought processes that bolster resilience during challenging situations.

Jennifer also helped start the MIT Resources for Easing Friction and Stress program. Student counselors in the program provide peer support, coaching, and mediation services. Peer-counseling programs such as the one at MIT have become much more common in the past decade and are currently part of counseling centers on about one-third of college and university campuses.

Some of the topics covered by Jennifer in her MIT programs are also on the agenda of the nonprofit organization Women in the Enterprise of Science and Technology (WEST) based in Cambridge, Massachusetts. WEST provides education, mentorship, networking, and information sharing for women to cultivate entrepreneurial thinking and creative risk taking. Its membership represents industries ranging from information technology and biopharmaceuticals to clean technology and environmental science. *Sandra Glucksmann* is the chair of the WEST Board of Directors and encourages interested women scientists and engineers to attend a panel discussion or networking event as an introduction to the organization.

Questions to consider: How attuned am I to gender or racial bias in my classroom, laboratory, or office? What more can my colleagues and I do to create an inclusive, welcoming environment? If I am encountering organizational barriers or bias, what resources at my institution should I contact? Is there a supportive senior person whose insights would be helpful?

FINDING MENTORS AND ADVISERS

The two mentors whose opinion *Shirley Tilghman* trusted most told her to keep going and not to let anyone tell her otherwise. Apriel Hodari and Beverly Hartline were mentors for *Renee Horton*, showing by their example how to be assertive and make people listen. Math professor George Rosenstein and his family gave *Wanda Austin* the support she couldn't get from her own family, who, without having had a college experience, couldn't understand the social and academic support she needed.

Physicist and General Lew Allen was a great role model for *Paul Nielsen* during the year when Paul was his aide. General Allen showed by example that a scientist or an engineer could get ahead in the air force and taught Paul about leadership. Professor Josh Fisher provided *Richard Lethin* a vision of how a computer should be architected, and so, following college graduation, Richard joined Fisher's start-up. Richard later founded his own company, which built on his mentor's vision but added his own technical understanding and values. Professor Jerrold Zacharias was the mentor *Rai Weiss* needed to help him find his way as an undergraduate at MIT. Lucia Rothman-Denes was a demanding adviser for *Sandra Glucksmann*, teaching her about the rigors of science and about how to live life to its fullest.

Each of the role models described in this book had his or her own role models. Some were influential teachers. Some were business executives or military leaders. Some were academic advisers. *Beth Reid* studied under *David Spergel*, whose thesis adviser was *William Press*, who studied with Richard Feynman and Kip Thorne. In a field based in apprenticeship learning, scientists gain credibility from their choice of adviser and his or her recommendation. *Rai Weiss* goes so far as to say that for a young scientist it is critical to have an influential mentor who believes in him or her.

One possible strategy for success as a graduate student is to find a prominent scientist for an adviser. When *Chris Loose* returned to graduate school, he knew that he wanted to work with MIT professor Robert Langer, a world expert in drug delivery, tissue engineering, and biomaterials as well as the founder of many biotech start-ups. Chris was interested in starting his own biotech company, so working with Langer gave him the expertise and contacts to do so, even while he was still a graduate student.

When *Stephon Alexander* studied physics at Brown, he had three prominent professors as advisers, two of whom (Leon Cooper and Mike Kosterlitz) were Nobel Prize winners. Stephon was enthusiastic and imaginative but not one of the better students in the department. However, his advisers saw in him a creative spark and encouraged him to work on some of the most difficult theoretical problems in neuroscience and string cosmology. As a graduate student, Stephon really needed that affirmation and says that if his advisers hadn't been so encouraging, he would have left school.

Jessica Seeliger chose prominent professor Steve Boxer as her adviser because he had a deep understanding of biology and the physical tools needed to investigate them. Jessica found the Boxer group to be an excellent training environment but Professor Boxer himself to be more of a hands-off adviser than she would have liked. He gave Jessica the independence to choose the direction of her research, and as a result she tended to wander between projects.

Many of the subjects commented on the degree to which their advisers provided direction and the amount of independence they had as graduate students. For students preferring more direction, possible strategies include choosing a promising early-career professor as an adviser or creating a team of advisers.

Jennifer Park knew she wanted close supervision as a graduate student in stem cell therapy. She chose Professor Song Li as an adviser and was his second graduate student. Because Professor Li had only a few students, he was able to give Jennifer the time and attention she needed. *Jane Wang* chose early-stage professor Dean Toste for her adviser in organic synthesis and was one of his first graduate students, but she also had a team of three faculty members advising her on a multidimensional project that required a range of expertise.

Some subjects made an effort early in their training to find the mentors they needed to get a start in their scientific training and career.

As a high school student, *Ethan Perlstein* read immunology journals while working as a technician in a biotechnology start-up. He sent letters to several prominent scientists with questions about their work. Ronald Germain at the NIH was kind enough to respond to these questions from a high school student, which led to a correspondence between them and then to an internship during the summer before Ethan started college. Ethan returned to the NIH after his freshman and junior years, and Germain became a mentor for him.

Kathy Gisser found her first mentor by visiting the Chemistry Department at Case Western Reserve. When as a high school senior Kathy went to the department looking for a chemistry project, Dr. Al Anderson was the first professor listed alphabetically in the directory. She called him up, and when she visited his office, he offered her a job. Assisting graduate students with research in Anderson's lab gave Kathy the confidence to succeed as a chemist.

As a high school senior, *Jessica Seeliger* had taken a thermodynamics course with Professor Norman Craig at Oberlin College. She contacted him and asked if she could work as an intern in his lab for the summer, and he agreed. Jessica assisted with spectroscopy research and operated and maintained expensive analytical equipment. The experience provided her with confidence in her experimental skills as well as the desire to continue in the area of physical chemistry in college.

As a college freshman, *Tamara Doering* took a biology class with Professor Michael Edidin and asked to be an intern in his lab for the summer. She was planning to work for him again the next summer, but when she found out he was going on sabbatical, she walked into the lab next door and asked Professor Saul Roseman for a job. She continued working for Professor Roseman throughout college, and the glycobiology research she did in his lab became central to her later experimental work.

Some subjects found their mentors by luck or happenstance. *Steve Fantone* calls such an event a "serendipitous encounter." He says his life has been full of such encounters, the most important of which was meeting his wife. For his professional career, the most important encounter occurred at 4:00 a.m. while he was standing in the ticket line at Logan Airport, when he met his future mentor and business partner, Sam Raymond, an encounter that opened up a host of unexpected opportunities.

Rai Weiss also found his mentor by chance. After flunking out of MIT, he was wandering the campus wondering what to do next, accidently walked into the laboratory of Jerrold Zacharias, and offered his assistance to one of the technicians there. Zacharias later taught Rai how to do highly precise physics experiments, helped him complete his degrees, and sponsored him for jobs.

Louis Pasteur famously noted that chance favors the prepared mind. He was referring to his own scientific discoveries and those of physicist Christian Oersted, but his observation also applies to how Steve and Rai found mentors.

Long before Steve met Sam Raymond, he had consciously made preparations to put himself in situations where fortuitous encounters might happen. For example, as an undergraduate he volunteered to set up the lecture hall for visiting Optical Society speakers, which gave him access to the visiting speakers. By graduation, Steve already had an extensive set of industry contacts and knew about research going on at all the companies in his field.

Similarly, preparation was the reason Rai ended up staying in the Zacharias laboratory after wandering in. Rai was skilled in electronics, having built and sold audio equipment as a high school student. That experience prepared him to recognize that the electronics in the lab needed improving, which led to a job in Professor Zacharias's research lab.

> Questions to consider: Do I already know whom I need to contact to get an internship or job? Who might I ask to be a mentor? What preparation, such as skill building, networking, or volunteering, can I do to maximize the likelihood of a serendipitous event?

WORK AND FAMILY BALANCE

Sandra Glucksmann and her husband, Charlie, had a daughter during Sandra's third year of graduate school. Sandra says that having a toddler didn't affect how long it took her to get through graduate school, but it did have a big effect on her work–life balance and her marriage. She and Charlie lost most of the liberties they had before their daughter was born. Charlie was working as a lawyer while Sandra was doing her thesis in molecular genetics. If Charlie had a trial or Sandra had to go back to the lab to do an experiment, they had to juggle their schedules. It wasn't easy, but they managed, and both were able to launch their careers.

For *Deirdre Olynick*, questions of work–life balance were major considerations for her beginning with her first job out of graduate school, three months after her son was born. She faced the realities of being a breastfeeding working mother. She and her husband chose a house and daycare close to Deirdre's workplace, which meant her husband had to make a very long commute. That unstable situation drove Deirdre to change jobs until she found a situation that felt right for her and her family.

Deirdre chose to work part-time job at Lawrence Berkeley National Lab to give her the flexibility to be more involved with her kids. She spent time volunteering at their school, which included helping with scientific activities and organizing school events. Once her children were older, Deirdre became a full-time employee at the Berkeley Lab. She subsequently finished an MBA, then changed fields, and currently has a full-time position directing a program in Global Health at the University of California–San Francisco.

Tamara Doering and her husband, Michael, who is also a professor, had to negotiate difficult work–family issues in order to pursue their careers. One choice they made was to delay having children until they were well-established professors. Some of Tamara's graduate students have asked her about the best time to start having a family. She tells them that having kids is amazing but that there is no one best time of life to have children. But Tamara also says that if she had had children during graduate school or her postdoc, she might not have done as well professionally. She believes that if she had been struggling in her early years to launch her research career while working the extra job of being a mother, she might not have succeeded in academia.

Jessica Seeliger and her husband, Markus, also chose to wait relatively late to have a child. Like Tamara, Jessica recognized the challenges of combining an academic career with raising children. She was concerned that it would be hard to give up time from her lab and teaching to care for a child and that being a mother would cause her to lose her focus. She reports happily that having a child has been a surprisingly joyful experience and a welcome contrast from her work. Her son, Carl, has helped her prioritize and has reacquainted her with playing and simple pleasures.

Daniel Theobald and his wife, Deborah, jointly created a company while starting a family. Within two years after starting their health-care software company, they had two children and a growing list of clients. Four years later they moved back to Boston, continued to grow the company, and had two more children. Daniel and Deborah found that having their own company was a great deal of responsibility but that it also provided them with more flexibility than employees generally have. For example, they could take their children into the office. Daniel and Deborah closely partnered to balance work and family while starting a business.

Jennifer Roberts and members of her committee in the Department of Electrical Engineering and Computer Science creatively reframed the issue of work–life balance when she was a graduate student at MIT. Before she arrived at MIT, the department would hold panels directed only at women students in which women professors with young children would talk about balancing career and family. So Jennifer helped organize panels that included both male and female parents from academia and industry and were directed at both women and men students. Jennifer believes a joint-parenting model empowers women to succeed professionally.

Both *Ellen Pawlikowski* and *Wanda Austin* would agree with Jennifer about the importance of shared parenting. They credit their spouses for the support that allowed them to succeed.

WORK AND OUTSIDE INTERESTS

Jami Valentine decided after an overly stressful time in graduate school that working at the US Patent Office was the perfect job for her. One of the reasons is that the Patent Office employees are encouraged to have a life outside of work. Jami takes advantage of her free time to volunteer and mentor students. Many role models have significant interests outside of science and engineering and have structured their careers to pursue these interests. One of the most common such interests is musical practice and performance.

Jennifer Park has a strong interest in music, and she even considered a career as a musician. She continued to perform classical music as a member of a wind ensemble. Both *Wendy Cieslak* and *Jessica Seeliger* are string musicians who enjoy playing classical chamber music. *Stephon Alexander* regularly performs as a jazz saxophonist. Music is central to Stephon's life, and over time he learned to combine his two passions, finding inspiration for his cosmological theories in his music.

As I shared in my essay, "Adapting and Creating," music is also central to my life. After a day of work, I love to play the piano. Playing music provides balance, an outlet for creativity and a way to connect to other musicians and listeners. Music is structured, just as science is structured, with standard notation and agreed-upon classical and jazz forms. And music allows me the freedom to explore, to improvise, and to make discoveries, as is true for scientists in the world.

As scientists or engineers considering a direction, the ways our career path interacts with our personal lives is at the heart of our choice. In their views, stories, and choices, many of the role models in this volume describe lives in which they balanced their interest in science and engineering with personal goals, such as time for family and outside interests.

Questions to consider: What personal commitments and interests do I hold dear? In making career choices, how can I balance my personal and professional lives in a way that honors both without feeling that I have sacrificed one for the other?